FROM CAPE COD TO THE BAY OF FUNDY

AN ENVIRONMENTAL ATLAS OF THE GULF OF MAINE

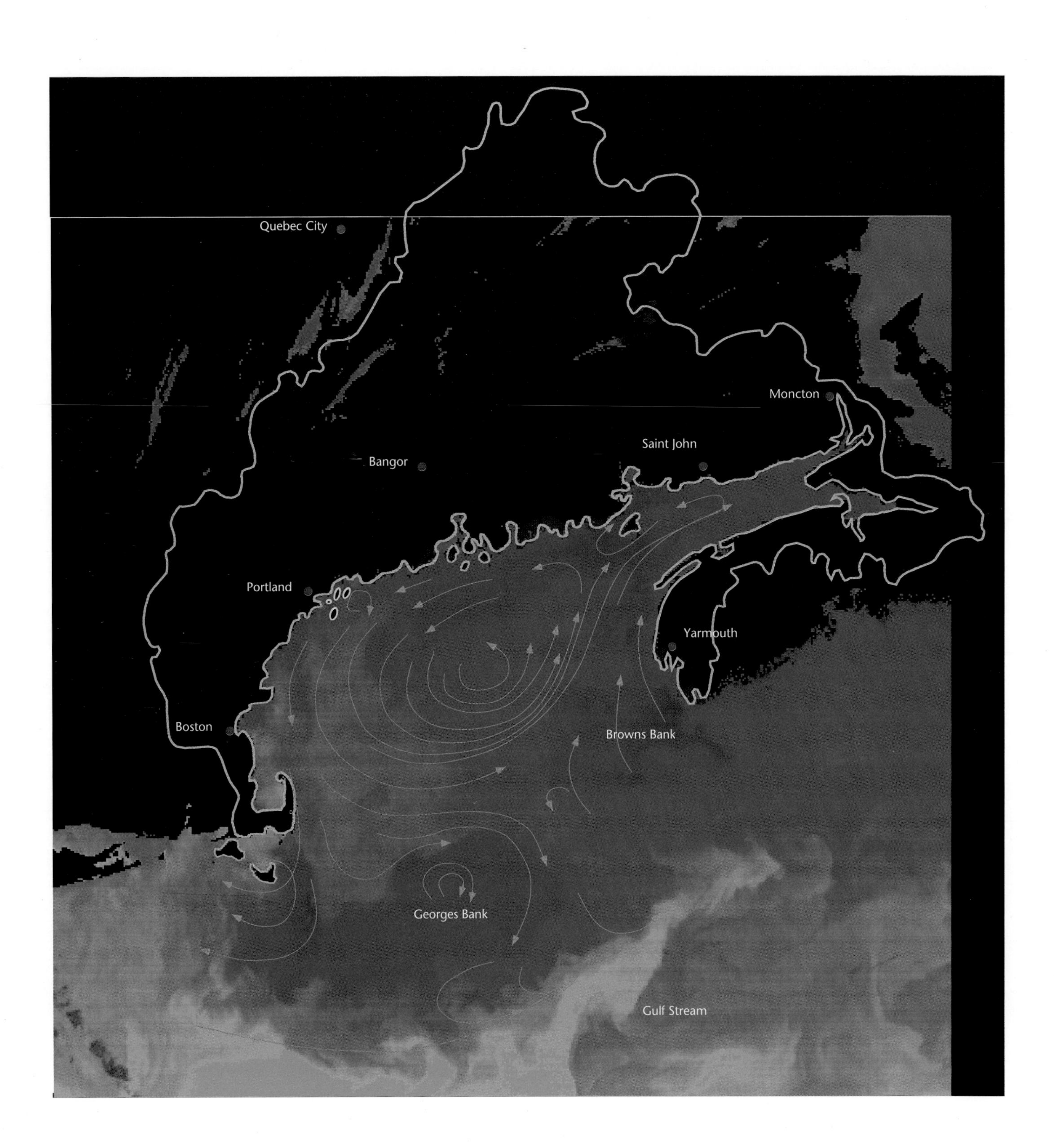

Quebec City
Moncton
Saint John
Bangor
Portland
Yarmouth
Boston
Browns Bank
Georges Bank
Gulf Stream

FROM CAPE COD TO THE BAY OF FUNDY

AN ENVIRONMENTAL ATLAS OF THE GULF OF MAINE

EDITED BY PHILIP W. CONKLING

THE ISLAND INSTITUTE

The MIT Press
Cambridge, Massachusetts
London, England

This book was set in Garamond 3 and Stone sans serif by Colotone Group and was printed and bound in the United States of America.

Library of Congress Cataloging-in-Publication Data

From Cape Cod to the Bay of Fundy: an environmental atlas of the Gulf of Maine/edited by Philip W. Conkling.

p. cm.

Includes bibliographical references and index.

ISBN 0-262-03227-9 (hc). —ISBN 0-262-53127-5 (pbk.)

1. Ecology—Maine, Gulf of. 2. Maine, Gulf of—Environmental conditions. 3. Physical geography—Maine, Gulf of. 4. Ecology—Maine, Gulf of—Remote sensing images. 5. Maine, Gulf of—Environmental conditions—Remote sensing images. 6. Physical geography—Maine, Gulf of—Remote sensing images. I. Conkling, Philip W.

QH105. M2F76 1995

574.5'2636'0916345—dc20 95-3319

CIP

To the next generation's stewards of the Gulf of Maine,

and most especially to

Timothy, Samuel, James, and Micah

CONTENTS

PART TWO: MOVING ASHORE

ILLUSTRATIONS

Appendix

ACKNOWLEDGMENTS

Six years ago Richard Podolsky and I developed an early version of a satellite imaging software program for interpreting ecological features among the thousands of islands with which the Island Institute was concerned. Podolsky was the Institute's first research director and brought with him a keen interest in ecology, computers, and high technology. He provided key leadership at the Institute during the development of the software application, GAIA (an acronym for Geographic Access, Image and Analysis), which was initially supported by a hardware grant from Apple Computer. With the assistance of land trusts on Islesboro and Vinalhaven, and with the help of Tom Goettel of the U.S. Fish and Wildlife Service, we assembled the Institute's initial archive of satellite imagery of the region. It was out of these early efforts that the Atlas grew.

Two years later Janet Campbell, a highly skilled remote sensing specialist then at Bigelow Laboratory for Ocean Sciences in Boothbay Harbor, Maine, initiated the Gaia Crossroads project, using the Institute's software to teach various terrestrial and ocean science concepts to schoolchildren. We were encouraged that experienced professionals like Campbell thought our software accessible enough to excite student interest, but none of us really knew how far kids would be able to push its uses until an additional 20 early color computers were donated to the Gaia Crossroads project by Apple Computer through the interest of Fred Silverman, who supervised the company's grants program.

Integrating satellite image projects completed by students with those of professionals would have remained just another untested idea if the A. K. Watson Charitable Trust had not decided to take a significant grantmaking risk. The early support of the A. K. Watson Charitable Trust and members of A. K. Watson's family, including Nancy Symington, Jane W. Stetson, Anne W. Bresnahan, Carolyn W. Morong, Arthur K. Watson, Jr., David J. Watson, and Stuart H. Watson, was crucial.

Through the efforts of Janet Campbell and her colleague, Cynthia Erickson, hundreds of students in scores of

Maine classrooms developed projects to be considered for publication in the Atlas. We appreciate the participation of the Gaia Crossroads pilot teachers from across coastal Maine, including Bob Dyer, Andy Vail, Dean Meggison, and Gary Seekins of Kennebunk Schools; Donna Weigel, Bonnie Dill, Lawrence Ryan, and Marylou Ryan of Wells/Ogunquit Schools; Katie Bauer, Frank Callanan, Philip Marcoux, and Margo Murphy of Cushing, Tenants Harbor, and Thomaston Schools; Scott Horr, Earl Edwards, and Joseph Salisbury of South Portland Schools; Jon Kerr and Miles Standish of Islesboro School; Jeff Crawford of North Haven School; Arden "Georgi" Thompson, Sue Lippert, Ralph Keys, and Diana Sommers of Wiscasset Schools; and Judy Dorr, Kathy Closson, Jill Clay, Sherrie Hersom, Bob Walters, and John Lunt of Boothbay Region Schools.

The time for organizing the writing and editing of the book would not have been possible without the generous support of Thomas J. Watson, Jr., who sadly did not live to see the finished work. Additional support came from the IBM Foundation's matching grants program. Emily Muir of Stonington also made a significant contribution toward the time for writing this book.

The conceptual approach to the book evolved out of long and frequent discussions with Spencer Apollonio, who nearly fifteen years ago authored a brief but significant book, The Gulf of Maine. The editing of the present volume involved a close collaboration with Richard Podolsky, and our regular editorial meetings benefited from the creative input of Cynthia Erickson of Bigelow Laboratory, Janet Campbell of the University of New Hampshire, John Anderson of the College of the Atlantic, and a host of Island Institute staff including Annette Naegel, David Platt, Peter Ralston (who contributed many of the photographs), Cynthia Bourgeault, Taylor Ongaro, Carol Smith (who chased endless details that occasionally threatened to engulf the project), and most especially Suzanne Meyer, who coordinated this complex undertaking with authors, photographers, contributing scientists, government officials, and Gaia Crossroads students and teachers.

Virtually all of the editing and a portion of the writing of the Atlas were completed in office space overlooking the edge of Penobscot Bay generously provided by Jan Taft. The preproduction work on many of the images was completed at the facilities of the Center for Creative Imaging with the help of Rusty Brace and Karen Brace.

Critical contributions to the printing and publication costs of the Atlas were underwritten with generous grants from the Trustees of Davis Conservation Foundation, Birch Cove Foundation, the Kendall Foundation, Bonnell Cove Foundation, Surdna Foundation, Inc., the Irving Foundation, and the Branta Foundation, and through the generous support of Maine Yankee, Inc.

The book caught the early interest of Madeline Sunley of the MIT Press, who skillfully and cheerfully steered it past the numerous shoals on which it could have foundered. Matthew Abbate is the book's dedicated copyeditor and Yasuyo Iguchi its talented designer.

All of the book's coauthors read carefully through many versions of the manuscript and made many helpful suggestions and corrections. Janice Harvey and her colleagues at the Conservation Council of New Brunswick, especially Inka Milewski, made many important contributions to the sections dealing with international aspects of the ecological issues presented in the book. We additionally wish to credit the Maine Scholar, in which portions of Lloyd Irland's chapter on the region's forests originally appeared.

Yet the manuscript last stopped on my desk, and any errors that were not ferreted out are finally my responsibility alone.

Philip Conkling

Aerial View of the Island World of Midcoast Maine

Credit: P. Ralston

PREFACE

NEW ENGLAND'S PROSPECT, 1994

George M. Woodwell

This book had its origins in the combination of the rapidly emergent technology of satellite imagery and the need to address the equally rapidly developing human onslaught that threatens to overwhelm every corner of the earth. Few of us are immune to the threat, which is at once general and remote and apparently beyond control, yet immediate and personal and crying for simple action in the common interest. Maine, northeasternmost of the United States, a frontier, rich in resources but one of the nation's most impoverished states, retains more potential than most for controlling its future. Suddenly, through an emergent technology, we have a new mirror in which to look at ourselves and perhaps find a different and less threatening course.

Only a short time ago, a generation or so, the world was different.

My father, a minister's son, lived in four Maine towns during his youth. In his teens he lived down on the coast in South Bristol. In those early years of the second decade of the century, South Bristol was a tiny coastal village isolated by geography and the independence and confidence of its citizens but dependent on wood for fuel, on boats, the sea, and coastal commerce. Only a few years later Harvey Gamage became famous for yachts and brought fame to the village. School for Father was Lincoln Academy in Damariscotta, fourteen miles up the Damariscotta River by steamboat. In winter, when the river froze and the steamboat could not run, he and his sister moved to live with the minister either in Damariscotta or across in Newcastle, I can't remember which. But Dad remembered it all, every detail, in full color, and he told it all to his young son in a thousand bedtime stories that etched themselves into memory as vividly as they must have lived in his: the woodpile, the grocery store, the automobiles of summer people, right down to his own speculations about the feasts of the earlier owners of that rich province who left shell heaps up and down the coast, remnants of what had to have been, in his mind at least, magnificent summer banquets joined by hundreds, equivalent to county fairs of our own time.

The minister's family, however, was poor; there was never food left over. The minister was the scholar in town and might well be the schoolteacher and hold several other offices. But the pay was low, if it existed at all. When "sho't" lobsters appeared on the doorstep, no one measured them. If life was hard, it was fun for that boy and his sister, and the telling of it made vivid experience for another generation. When I finally saw South Bristol on a summer trip as a youth, it could never live up in one afternoon to the distilled excitement of youthful years spent exploring life with a bright and venturesome sister in that ancient Indian banquet ground.

South Bristol and the river were always a large but highly intense and personal world to Father, far away from any other place. When he went off to Dartmouth, he left in the fall to return in the spring to walk, carrying his suitcase, the fourteen miles down the peninsula. Once, years later when he came home from the war on Christmas Eve to visit a mother ill with the terrible influenza of 1918, he and his brother caught a ride down to East Boothbay, across the river, borrowed a skiff, and rowed in the December cold and dark across to South Bristol and home. His close friend, one of the three from that village who went off to war, never came home, killed in France the day before the Armistice. The world was also small and personal and difficult.

Recently, research in Maine took me to Bangor in a small plane on a glorious fall day. We flew down the coast from Cape Cod, over the Isles of Shoals, past that forbidding rocky shoal called Boone Island, across Casco Bay, past Halfway Rock and eastward. There, suddenly, in one glance below lay that entire fairyland from Pemaquid Point and Christmas Cove to Damariscotta. It could fit into one quick look, one eyeful, land and water and history, Indians, ice, and steamboat, all in the eye and mind at once. A youth spent on the coast of Maine in the early part of the century and recorded as vivid history in the mind of a son, visible now from three thousand feet through the window of a machine that, with its congeners, has in one lifetime reduced not only that region but the earth as a whole to one eyeful.

Father never shared in that experience, now routine and encoded in a thousand ways, that makes it possible for us to look at once at New England from Dartmouth College to Lubec, or at the entire Gulf of Maine, in a single glance. The imagery (they are no longer called pictures because the information has been so extensively refined as to be far beyond a mere picture) comes to us from airplanes and from satellites and gives us an instant view of a whole we could not easily comprehend just a few years ago. And it comes often enough that we can keep track of changes in the land or water globally, twice a day in some instances. We have a new frontier in science, a thousand new ways of looking at the world and measuring it almost minute by minute. And, if we wonder where we are, anywhere in the world, from a sailboat in the fog off Pemaquid Point to a bend in the Yenisey of central Siberia, a satellite-based system with a hand-carried device no larger than a telephone will give an instantaneous position within a few feet. Other devices are accurate enough to measure the spreading, year by year, of the Atlantic as the continents continue their 300-million-year drift.

The airplane, the satellite, the "imagery" that replaces pictures are all marvels of technology. They convey command of new resources, information, time, air, and land. This new power, almost unearthly and godlike, a product of a fossil fuel–powered technology, seems to belie any vestige of a fundamental dependence on life apart from our own. How can it be that our ability to collect data such as those from the images recorded in this volume depends on the living systems that we see in the imagery as forests or oceanic plankton? We have to step back still further to realize that the view of the earth we have here is of an earth that is not working properly: it is losing its capacity to support life, even as we gain the capacity to see that life in new ways.

Changes in the atmosphere are causing a warming of the earth that promises large changes in the structure of nature, changes in the ecosystems that contain, coded in the genes of

Southern New England from Space

The region from Portsmouth Harbor and Isles of Shoals (top) to Martha's Vineyard and Nantucket (bottom right). The Merrimack River enters the ocean near the top to create Plum Island Beach and Wildlife Refuge. Yellow areas are major salt marshes. Major urban areas (gray) to the northwest of Boston include Lawrence, Lowell, Nashua, and Manchester, all of which are on the Merrimack River. Buzzards Bay, rimmed by the Elizabeth Islands (bottom center), is relatively undeveloped except at New Bedford. Narragansett Bay is heavily developed around most of its periphery. Dark areas in the image are major areas of wetlands. *Credit: Space Shots*

their plants and animals, the information required to keep the earth running as the habitat of all life. Other changes are causing an increase in the intensity of ultraviolet radiation at the surface of the earth, a change that is clearly deleterious but whose effects will probably be subtle, realized only over time as systematic but unspecific impoverishment of those same systems—and as an increase in skin cancers and other even more subtle effects on people. The changes come as a product of the fossil fuel age, as much a product of the technological age as the aircraft and the satellites and the smelted aluminum and steel and the power that carried my colleagues and our equipment and me down the coast that bright fall day.

If we scan these images carefully we can identify the process, the creeping necrosis of pavement, shopping malls, gravel pits, and subdivisions. If we look to the north in Maine, we can see the unorganized townships, most corporately owned, defined on the landscape in six-mile squares by the intensity of harvest of but one of a forest's products, timber. The winter images are definitive: the more heavily cut are bright, white in fact with snow. The coniferous forests, where they survive, are black. The difference is that sharp, white and black, cold and warm, nonforest and forest, skinned for profit now or managed for the longer term. The issue here is timber, pulp, profit. No thought here for the world as a whole, for the information pool that keeps the forest renewable century after century and the human habitat working to support all life. No thought here for the public interest in intact forest as a stabilizing element in the human habitat, as a source of clean water, as a reservoir of life, as a resource in itself in support of all. Yet here it is, the world in miniature. The whole issue of a world that works versus a world that does not is embraced in a single landscape, now visible to all. The excesses of exploitation in Maine are assumed to be accommodated by New Hampshire, or Massachusetts or Canada or the Amazon basin, or by an intact landscape in some other place. But, if we were to look, as we can now, we discover that there is no other place free from the twentieth-century necrosis, now obvious in images such as these.

That's the news: the Gulf of Maine region, proud of its heritage of independence, intellectual and individual and collective, has, in one generation, gone from the reality of that independence to the reality of dependence on resources elsewhere that do not exist. If that's the only message of this book, it is a message well worth having. If the message brings the question of how to turn that circumstance around, how to work out patterns of land use that are, in fact, self-contained in the centuries-old traditions of small towns, we would have a contemporary model for the world.

Why not start now?

PART ONE: EDGES OF THE SEA

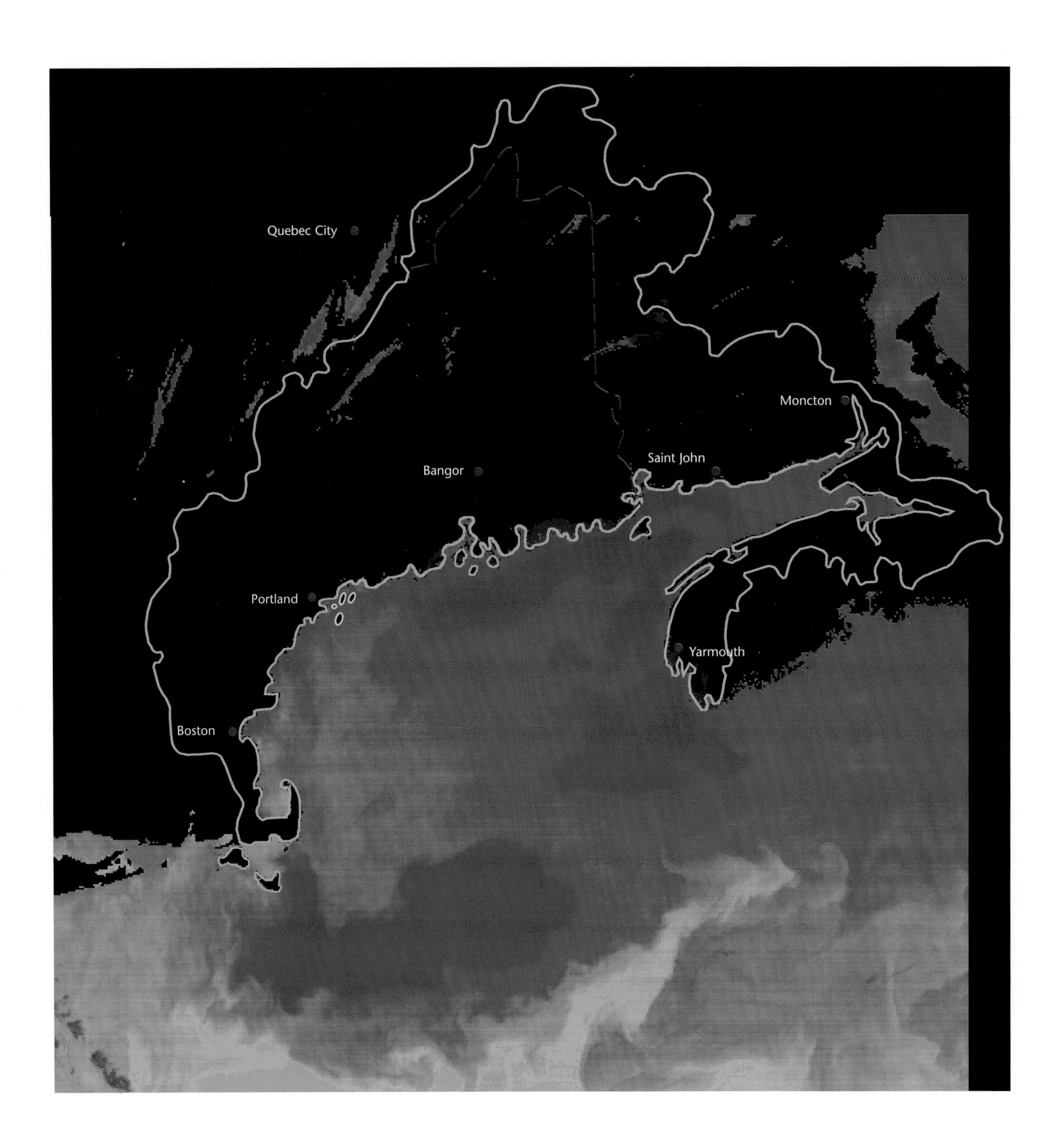

Quebec City
Moncton
Saint John
Bangor
Portland
Yarmouth
Boston

1

AN INTERNATIONAL COMMONS

Philip W. Conkling

1.1 Gulf of Maine International Watershed Image

This infrared satellite image shows subtle differences in water temperatures during the summer in the Gulf of Maine. The warmer waters of the Gulf Stream in two shades of green are visible at the bottom of the image. Cold waters in dark blue appear over Georges Bank and the Scotian Shelf. North of the Scotian Shelf, a tongue of cold water is visible to the west of Yarmouth that trends along the eastern coastline of Maine. Waters in the western Gulf of Maine between Portland and Cape Cod are distinctly warmer. *Credit: image processed by T. Aarup, Bigelow Laboratory for Ocean Sciences; data furnished by G. Feldman, NASA Goddard Space Flight Center; graphics: S. Meyer; image enhancement: P. Conkling, T. Ongaro*

In the northeastern corner of North America, a large international watershed empties into a semienclosed body of water called the Gulf of Maine (fig. 1.1). For more than four millennia this ecosystem has cradled a complex forest of hardwoods and softwoods, produced an incomparable supply of fish and shellfish, and provided an important basis of international trade and wealth for communities on both sides of the North Atlantic. Ever since glacial meltwaters poured into the rolling coastal lowland, raising the level of the sea to create the Gulf of Maine, people have been a part of this ecosystem. A deep connection with the region's rivers, forests, and bays is as ingrained in the maritime and forest communities that stretch across its wide expanse as in the cities whose people depend on its resources for commerce, recreation, and spiritual renewal.

For virtually all of our history, these resources have seemed inexhaustible. However, an ever-increasing number of citizens across state and international boundaries have begun to recognize that the natural resources that have sustained human life for centuries around the Gulf are preciously finite. Canada has recently suspended all fishing for cod and other groundfish in the waters off its Atlantic provinces, while in the United States federal regulators have closed Georges Bank for the first time ever to protect depleted cod, haddock, and flounder stocks. As the catch of groundfish has plummeted, so has the economic well-being of tens of thousands of fishermen in hundreds of communities around the rim of the Gulf of Maine. Meanwhile millions of seafood consumers in both countries are having to shift to imported products. For more than two centuries the northeastern United States and Atlantic Canada have been one of the world's leading seafood-exporting regions; but during the last decade we have been transformed into one of the world's largest seafood importers, partly as a result of declining wild supplies and partly as a result of our increasing appetite for healthy sources of protein that fish and shellfish provide.

Even the lobster, another indelible symbol of the coastal region, appears to be in jeopardy. In 1993 the State of Maine issued a health warning advising children and pregnant women not to eat lobster tomalley due to contamination by dioxin, which is a by-product of the chlorine used in bleaching paper and in disinfecting sewage effluent from hundreds of town treatment plants (fig. 1.2). Meanwhile in the southern part of the Gulf of Maine, blooms of toxic red tides have developed in previously pristine waters.

1.2 Larval and Juvenile Lobsters
The first surface-swimming stage of a larval lobster, and a juvenile lobster after it settles on the bottom following several molts. Larval and juvenile lobsters, like all fish and shellfish, are highly sensitive to small concentrations of toxins in the water column, including chlorine. The State of Maine has recently issued health warnings advising pregnant women and children not to eat lobster tomalley, the lobster's liverlike organ, due to potentially unhealthful levels of dioxin. *Credit: K. Bray (after Herrick)*

Away from the coast, in both Canada and the United States, citizens are increasingly concerned about the scale and apparent destructiveness of modern forest harvesting operations. Huge feller-buncher machines lumber across industrial forest landscapes, clipping whole trees and processing them into chips for energy and pulp. New wilderness reserves have been proposed for large expanses of these northern forests on both sides of the border, but currently only a small fraction of the north woods is permanently set aside and restricted from exploitation.

We tend to think of this region as largely rural and therefore undeveloped, but this is inaccurate. Across the north from New Hampshire to New Brunswick is a vast industrial forest intricately laced with private logging roads. The effluents of Boston and Saint John at either end of the Gulf of Maine have effects far beyond their harbors. The Gulf region is the third most densely populated coastal region in the United States, and the population here is still inexorably growing. The signs of stress on natural systems are omnipresent and troubling.

What Is an Ecosystem?

The problems that confront us are symptomatic of our failure to think of the region as an interconnected whole, as an ecosystem. We are like the proverbial blind men describing the parts of an elephant.

Today you cannot pick up a newspaper or magazine without reading about "ecosystems," but does anyone really know what one looks like? Although nobody can really say they've "seen" an ecosystem, which is after all an abstraction of the human mind, most of us have a vague sense of what one is. We all remember the freshwater aquarium of yesterday's science class, where aquatic plants produced oxygen for guppies and tadpoles which, in turn, produced carbon dioxide for plant uptake. Visualizing an ecosystem, particularly one as large as the Gulf of Maine, is a lot more difficult, although the principles of interconnectedness are the same.

A few years ago, a survey vessel out on Georges Bank collected a new species for that locality, a tiny single-celled creature called *Alexandrium tamarense*. This species is actually well known and well studied in this region because it can erupt in a red tide bloom, causing paralytic shellfish poisoning in people who have eaten mussels or clams from an affected area. But in the past half-century of intensive scientific surveys, *A. tamarense* had never before been collected far offshore on Georges Bank. Although red tides are a naturally occurring phenomenon in many oceans of the world, as in the Gulf of Maine, there is little doubt among scientists that the incidence of red tides is increasing, and many believe this is related to increased nutrient loadings from human activities in the coastal fringe where we increasingly concentrate our activities.

Here in the Gulf of Maine, satellite images show the mouth of the Kennebec River, which drains a vast river system in Maine and New Hampshire whose banks are dotted with a half-dozen large pulp and paper mills, to be one of the prime progenitors of red tide blooms. From space, this river during a spring freshet looks like a huge outfall pipe (see fig. 11.9). The principle of interconnectedness means that the engineer whose job it is to control the contents of a sewage treatment plant or paper mill effluent in northern New Hampshire can affect the livelihood of a fisherman hundreds of miles away out to sea, even though their worlds are largely unknown to each other.

This principle of interconnectedness also has application for the stocks of dozens of species of fish that move freely back and forth across the international boundary between Maine and Canada. The regulations governing fishing on one side of the border obviously cannot work without a common understanding of their status and some form of information sharing, if not joint management. No official recognition of this obvious fact has yet penetrated our political systems. The proposal to harness the immense tidal power in the Minas Basin of the Bay of Fundy, if implemented, could change the tides in the entire Gulf and raise the sea level in downtown Boston. The steady stream of pollutants that we daily flush into our rivers slowly circulate in the great tidal gyre of the Gulf of Maine. Out of sight is not out of the system.

There is, however, a new tool, or more accurately a set of tools—cameras and sensors providing us with views from above, from orbiting satellites, shuttles, and aircraft—that for the first time are beginning to help us visualize the complex interrelationships in the natural environment that have thus far eluded our vision and imagination. In some cases, these cameras and lenses present us with images of features that have not been apparent to the human eye—how water moves through a

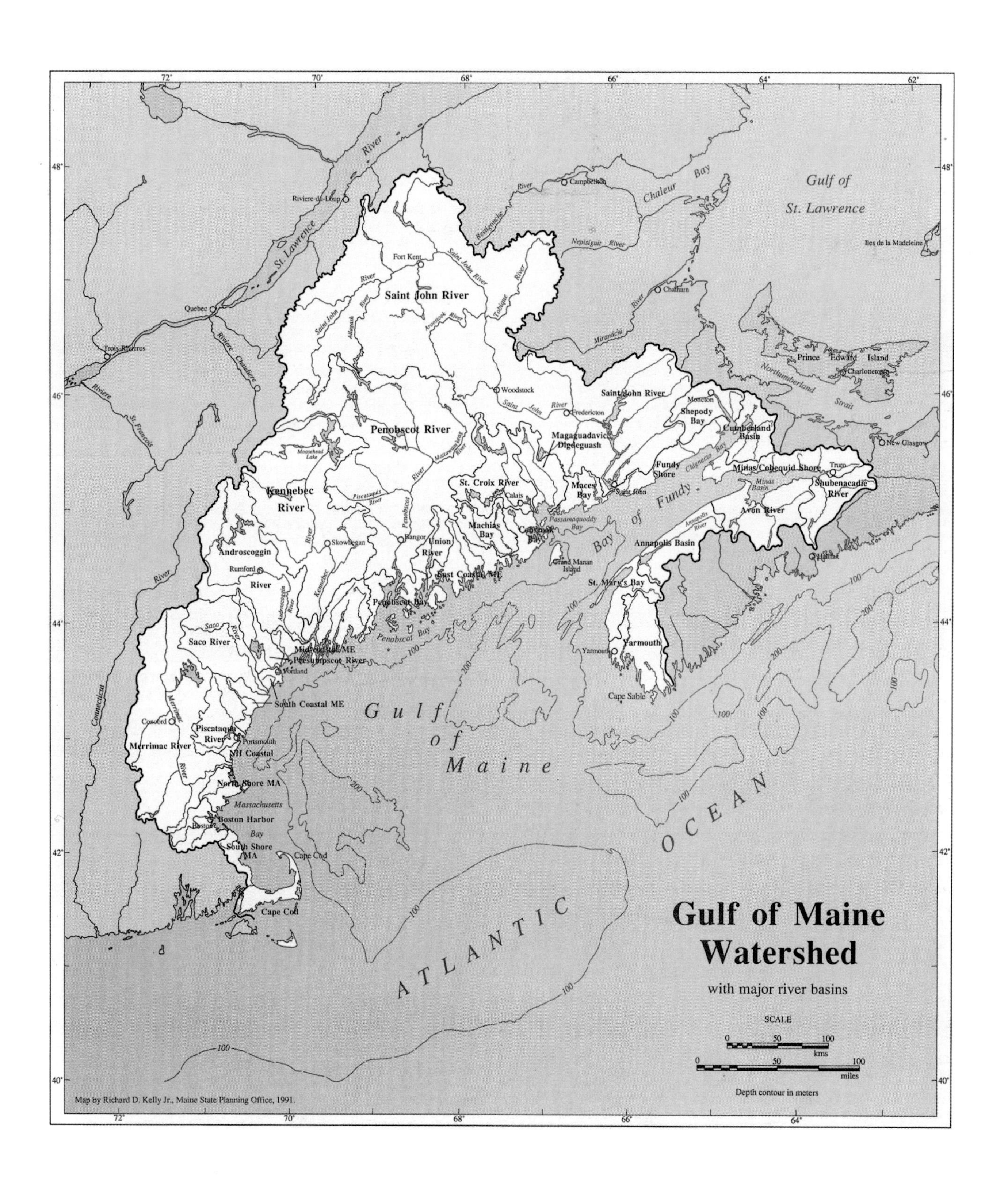
Gulf of Maine Watershed
with major river basins
SCALE
0 50 100
kms
0 50 100
miles
Depth contour in meters
Map by Richard D. Kelly Jr., Maine State Planning Office, 1991.
Gulf of St. Lawrence
Chaleur Bay
Bay of Fundy
Gulf of Maine
ATLANTIC OCEAN
Saint John River
Penobscot River
Kennebec River
Androscoggin River
Saco River
Merrimac River
Piscataqua River
St. Croix River
Machias Bay
Union River
East Coastal ME
Penobscot Bay
Midcoastal ME
Presumpscot River
South Coastal ME
NH Coastal
North Shore MA
Boston Harbor
South Shore MA
Cape Cod
Magaguadavic Digdeguash
Maces Bay
Fundy Shore
Shepody Bay
Cumberland Basin
Minas/Cobequid Shore
Shubenacadie River
Avon River
Annapolis Basin
St. Mary's Bay
Yarmouth
Cape Sable
Prince Edward Island
Northumberland Strait
Quebec
Trois-Rivieres
Riviere-du-Loup
Fort Kent
Woodstock
Fredericton
Moncton
Saint John
Calais
Bangor
Skowhegan
Rumford
Portland
Concord
Portsmouth
Boston
Halifax
Truro
New Glasgow
Charlottetown
Chatham
Campbellton
Iles de la Madeleine
Grand Manan Island
Massachusetts Bay
Moosehead Lake
Passamaquoddy Bay
Minas Basin
Chignecto Bay
Connecticut
St. Lawrence River
Riviere Chaudiere
Riviere St. Francois
Restigouche River
Nepisiguit River
Miramichi River
Tobique River
Aroostook River
Allagash River
Mattawamkeag River
Piscataquis River
Annapolis River

1.3 Gulf of Maine Watershed Map

The watershed and major river basins, produced by the Gulf of Maine Council on the Marine Environment. *Credit: R. D. Kelly, Jr., Maine State Planning Office*

forested wetland, for instance, or where there are concentrations of tiny single-celled algae in the ocean called *phytoplankton*. Although these tools cannot detect all the information we need to monitor in our environment, such as where toxic contaminants may be accumulating, they can "see" how currents interact with river discharges and instruct us where we ought to be looking. But more important, these tools have the potential, if wisely used, of transforming our view of the world around us and our understanding of our relationship to the natural world.

Beyond Political Boundaries

In 1989, when the governors of three American states and the premiers of two Canadian provinces created the Gulf of Maine Council on the Marine Environment, one of their first accomplishments was to produce a watershed map of the region, outlining the boundaries of the international watershed we share (fig. 1.3). This simple map is one of the few representations of any international region that does not impose the artificial political boundaries we have come to expect when viewing a part of the globe. Instead it presents us with a new view of the ecosystem we inhabit with people from different states and nations; this map reflects the basic ecological reality that we share an international commons.

We tend to conceive of political boundaries—those artificial lines on maps that separate us into towns, school districts, states, provinces, and countries—as defining the basic units of human society. Without question these units exert a major influence in our daily lives. But as we continue our inexorable proliferation across the face of the globe, with ever greater impacts on the natural environment, we also need to understand the complex interconnectedness of our environment within another set of boundaries—ecological boundaries—if we are to have any hope of wisely using the vast treasure of natural resources that the earth has offered up.

A watershed is in many respects a fundamental ecological unit we are only beginning to appreciate. It is often difficult to accurately define the boundaries of any ecosystem: the precise line on the ground that divides a desert from a grassland or a forest from an alpine region. But the path that water takes from the top of a ridge downhill to the sea is a defining feature, because water flows in only one direction, sustaining all forms of life on its way to the sea. It is precisely at this aquatic interface, the *estuaries* where fresh water mixes with the sea, that we find our most serious problems. Except for the offshore banks, coastal estuaries are the Gulf's most productive environments and are the prime spawning and nursery habitat for a stunning array of fish and shellfish. But estuaries are also where a river's contaminant load picked up along its entire route is deposited.

To understand what is going on in the Gulf of Maine requires an understanding of how different parts of this vast international watershed are linked. That is the aim of this book. Like any atlas, it will look systematically at various features of this region—the topography, rock formations, ocean currents, rivers, and forests. But we will also be asking how the parts fit together; how changes or disruptions in one area affect other areas.

A Sea within a Sea

Out in the ocean, drawing an ecological boundary becomes very difficult, often impossible, because systems there are so vast and there are few natural barriers between them. But here in the Gulf of Maine, the watershed principle holds even below the surface of the sea. The high hills and underwater ridges of Georges and Browns banks that rise up offshore create a unique marine

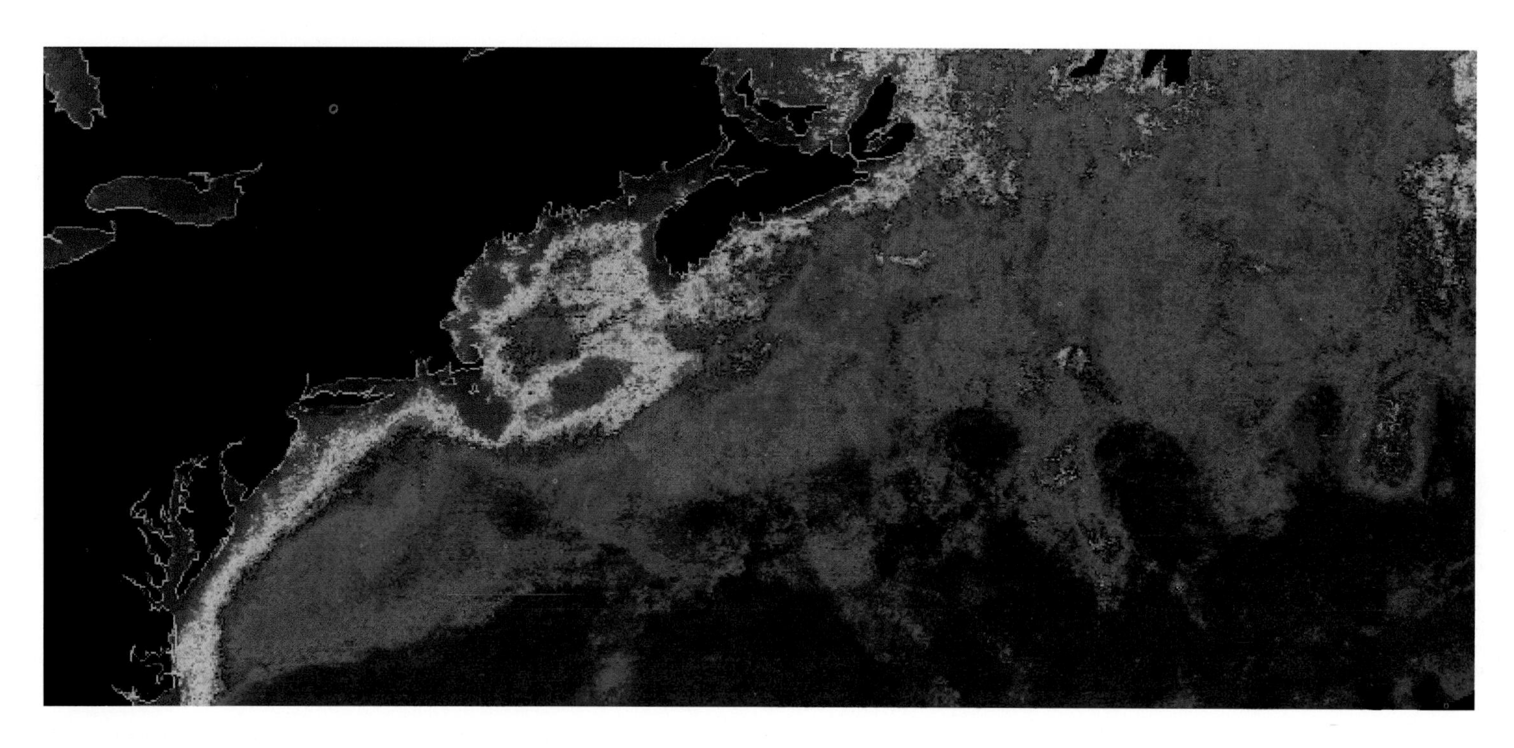

1.4 Sea within a Sea

Chlorophyll distribution from phytoplankton blooms in the North Atlantic, showing areas of high productivity (in red) along the entire continental shelf of North America. Note how the open ocean waters show the lowest concentrations of chlorophyll (in blue and violet). Note also how the distribution of enriched productivity appears around the rim of the Gulf of Maine to Georges Bank, showing it as a "sea within a sea." *Credit: G. Feldman, NASA Goddard Space Flight Center*

boundary between the Gulf of Maine and the Atlantic Ocean. These underwater topographic features, in combination with the lunar pull of the tides, create a more nearly enclosed circulation system than exists off most shores of the world. It is the edge of this circulatory gyre that defines the seaward boundary of the Gulf of Maine watershed and that underlies the world-class marine productivity with which the region is blessed. The Gulf of Maine is distinct from the Atlantic, an ecologically separate sea within a sea (fig. 1.4).

Another defining feature of the offshore portion of this international watershed results from the temperature of its waters. As anyone who has so much as dipped a toe in the Gulf of Maine can attest, swimming here can be a character-building experience. The waters that enter the Gulf of Maine are cold even in the summer, and this has a dramatic effect on all marine and shoreline life found in the region. The relation between temperature and biological productivity in water is almost the opposite of that on land. By and large, the further south you go on land the more productive the terrestrial ecosystems become: an acre of tropical rain forest supports more biological activity than an acre of arctic tundra. But in the water the reverse is true: the further away from the equator you go, the more productive, on average, the marine ecosystems are. To begin with, cold waters are capable of holding more dissolved gases such as oxygen and carbon dioxide, the fundamental building blocks of all life. Even more important for cold-water productivity is the phenomenon oceanographers call winter turnover. As northern waters cool in the fall and winter, they become more dense and sink. This deep convective mixing brings deep water and nutrients to the surface. In the spring, the waters stabilize and the surface layers explode in an early bloom of phytoplankton, months ahead of the terrestrial spring bloom.

The islands of the Gulf of Maine are another defining feature. An archipelago of over 5,000 islands rings the Gulf, creating immense expanses of subtidal habitat, past which surge powerful tides that stir these nearshore waters into a froth and bloom of biological wonder.

Because the Gulf of Maine and its islands are embedded in the temperate zone, the heat of the sun throughout the long days of summer and well into autumn warms the waters to extend the growing season, especially along the coast. This is the reason the shores and islands of the Gulf support so many plants and animals with geographic ranges that terminate here. Just as many northern species find their southern limit in the Gulf of Maine, numerous southern species have their northern limit here, resulting in a complex and rich assemblage of terrestrial and marine plants and animals that ecologists call *biodiversity.* Among birds, for example, the more southerly distributed glossy ibis, least tern, and laughing gull reach north as far as the Gulf, while more northerly species such as the Atlantic puffin, Leach's storm petrel, and black guillemot reach this far south.

Terrestrial diversity can pale beside marine biodiversity, especially in some of the habitats of the bottom of the Gulf of Maine where over 1,600 species of *benthic* or bottom-dwelling organisms have been described. In the summer, the heat of the sun stabilizes the waters over offshore basins, isolating the planktonic organisms from their deep-water nutrient source. But nearshore banks, ledges, and island shores remain productive throughout the summer because the vigorous stirring of the tides keeps phytoplankton richly supplied with nutrients in the upper sunlit layers where marine biodiversity reaches a maximum.

Our Backyard from Space

The authors, photographers, scientists, and computer engineers who contributed to this volume were all given the task of describing, either in word or image, some part of the Gulf of Maine with which they are familiar in ecological terms. Through these views and voices we have tried to integrate many different fields of inquiry. Because this work connects different scales and times, and because it spans natural regions irrespective of borders, we believe it is inherently ecological.

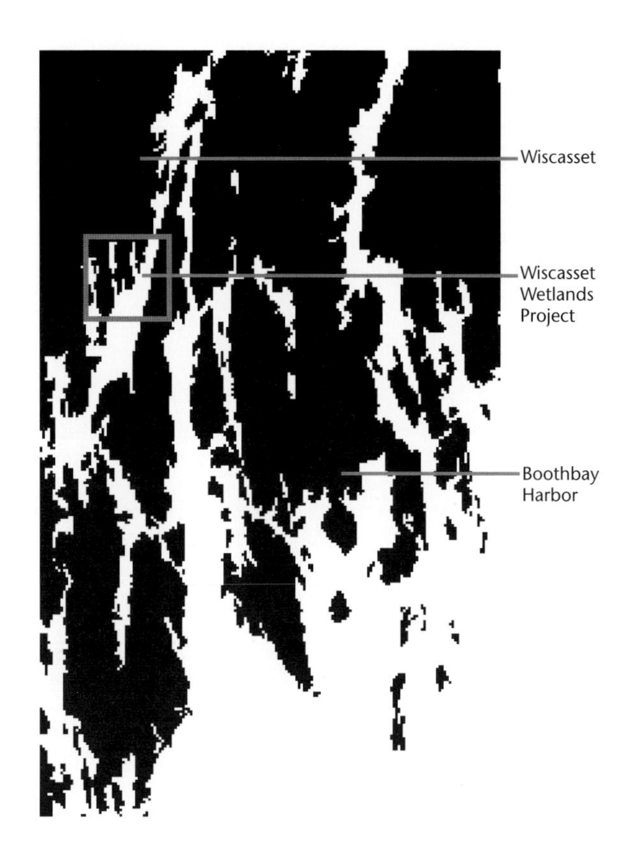

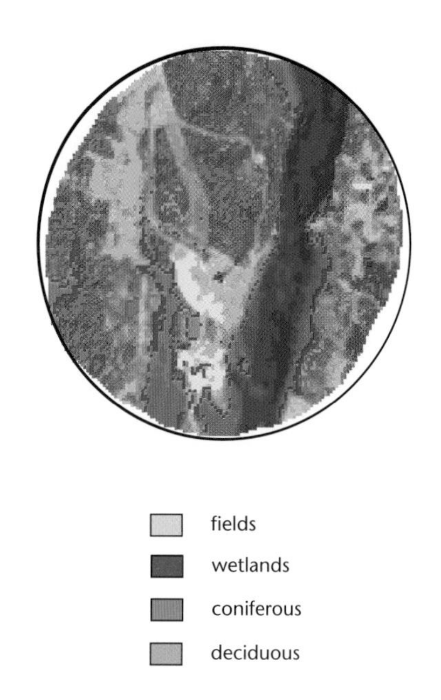

1.5 Wiscasset Wetlands

Students from the Wiscasset Elementary School in midcoast Maine had been theme-mapping their town using satellite imagery when Maine Yankee applied to the town planning board to construct a site for storing low-level nuclear waste at its plant facility on the Back River, south of town. The class was invited to make a presentation of their map to the planning board. In the words of their teacher, "The actual project we presented was one I compiled from all the student information that we had available and focused on wetlands, trees, and fields, because we wanted to include only habitats that we were absolutely certain about. Even so, the wetland pattern was startlingly clear. Our presentation came after a hydrogeologist spoke about what wetlands were and maybe where they were. These wetlands could be passed up as one-time extraordinary events until we showed our May 1988 image. My students and I didn't beat any drums or shout because we didn't have to. The neutral satellite image showed the proposed dump site quite clearly in the midst of wetlands." *Credit: L. Beerits, K. Bruce, and R. Lear, grade 3, Wiscasset Primary School; teacher, G. Thompson; courtesy of Gaia Crossroads Project, Bigelow Laboratory for Ocean Sciences*

These images and narratives, we hope, will serve as simple and comprehensible illustrations of the complex relationships between forests and rivers, and between freshwater rivers and saltwater spawning grounds of such species as lobsters, cod, and haddock. By looking carefully at these images, we can begin to visualize how weather patterns constantly reshape the borders of the southern rim and beaches of the Gulf, or how an unexpected current in the ocean can sweep in over offshore banks and affect fish larval recruitment, or how temperature fluctuations between the eastern and western parts of the Gulf can influence the catch of lobsters. In the past, natural occurrences such as these could be studied at the microscopic level by biologists, but they were imperceptible to most of the rest of us (although highline bluefin tuna fishermen have known for some years that downloading real-time satellite imagery showing temperature fronts can substantially improve their ability to locate that most economically valuable marine life to travel through the Gulf).

Some of the themes of this book are increasingly obvious and have obvious lessons: as sea level inexorably rises along the western edge of the Gulf of Maine shoreline, the costs of trying to hold back the sea will also rise inexorably; if we continue to ignore the subtle but cumulative addition of contaminants in nearshore environments from the effluents (especially when chlorinated) of large industrial and municipal outfall pipes, then immensely valuable fish and shellfish habitat will be permanently degraded.

But many other important relationships are not so obvious and await thoughtful observation and careful description. Ninety-five percent of the satellite images this country collected between 1973 and 1993 have never once been seen by human eyes, because expensive mainframe computers and a great deal of technical training were necessary to view them. Now powerful new desktop computers and easy-to-use software mean there is untold opportunity for each of us to gain new insights into ecological relationships in our individual and collective backyards.

We hope this book serves to demonstrate the opportunity earth imagery offers to monitor our changing environment, and encourages us to relate these observations back to the larger region—the ecosystem—that we inhabit. The most comforting thought is that the technical capacity to visualize how our corner of the world works is now within reach of all of us, from city to hinterland, from Atlantic Canada to rural New England, and from the region's most sophisticated scientific laboratories to the primary school children in Wiscasset, Maine, whose satellite maps were presented to a town planning board considering whether to approve a low-level nuclear dump (fig. 1.5).

The implications of understanding how the resources in our backyards relate to the larger environment and to planning a better future are immense. Viewed from space, we can see the vastness of our backyard—the Gulf of Maine watershed—at a glance. And seeing, perhaps for the first time, its natural boundedness, we can think of the region as a series of richly integrated habitat islands, separate islandlike watersheds that ultimately commingle in one vast watershed, or ecosystem, which is itself like an island in a surrounding sea and all a part of this island earth.

Perhaps these views will lead to a better understanding of the likely effects of many of modern civilization's present activities. Without a sustained commitment among each of us in our day-to-day life to see ourselves and our activities as part of a larger system to be passed along to future generations, the future is grim. Ultimately, it is easier to think globally than to act locally; but it is in our backyards that we meet the enemy, who more often than not, as Pogo has pointed out, is us.

Satellites: Tiny Specks of Light

Janet W. Campbell, Cynthia B. Erickson,
Philip W. Conkling, and Richard Podolsky

On dark clear nights, you can see them: tiny specks of light moving slowly through the sky. They appear like stars, but, unlike the fixed constellations in the celestial sphere, these stars move from horizon to horizon in the span of a few minutes. They are the heralds of a silent revolution that has occurred over the past few decades They are not stars, of course, but rather satellites that orbit the earth every 100 minutes or so, carrying cameras and sensors. They are charting the earth's spheres—our air, water, ice, and biological regions, which scientists call our atmosphere, hydrosphere, cryosphere, and biosphere, respectively. These satellites are components of the new technology of remote sensing.

Until the advent of hot-air balloons in the last century, the only way for humans to get a bird's-eye view of the earth was from the top of a mountain. With the invention of cameras in the mid-1800s, people began to make aerial photographs from balloons. One of the first balloon photographs was of Boston, Massachusetts, taken in 1860 from 1,200 feet above the city. A particularly intriguing photograph covering several square miles of the 1906 San Francisco earthquake and fire was taken from an array of 17 kites moored to a boat anchored in San Francisco Bay. Perhaps the most famous image of the earth from space is the "blue marble" photograph of the earth suspended in space that was taken by Apollo 17 astronauts on their journey to the moon in December 1972.

In much the same way that Seurat's pointillist paintings are actually composed of a multitude of tiny colored dots, digital satellite images are ultimately composed of millions of tiny visual subunits called *pixels* (short for "picture elements"). When viewed directly on a computer screen, pixels appear as different colored "tiles" in a mosaic arranged horizontally and vertically on the screen or monitor. Each pixel corresponds to a particular location on the earth, which is actually a square area of finite dimensions rather than a single point (fig. 1.6). The pixel is colored or shaded by the computer operator according to the data acquired by the satellite for that particular location. The degree of realism involved in the final image and the selection of colors used to represent the data are controlled by the computer operator, so, in a sense, each image is an artistic rendering of scientific data.

Each pixel represents a radiometric manifestation of a portion of the earth's surface, that is, the amount of radiation (visible light, infrared, ultraviolet, or thermal, depending on the remote sensing system used) being reflected by the particular land area displayed. Similar pixels have a similar radiometric manifestation, or what is referred to as a reflectance signature. The technology relies on two premises: first, that different pixels have unique signatures, and second, that their signatures are consistent for a given class of pixels regardless of their location in the image. Thus, the color assignments—whether realistically replicating what the human eye would actually see, or more abstract (because the data include information from the infrared or other regions of the electromagnetic spectrum that the human eye cannot see)—can be used to identify features at a glance. Important surface features of the earth can thus be identified, ranging from the density of microscopic plankton on the surface of the sea to the distribution of wetlands and the extent of deforestation and habitat loss. (For a more detailed description of how this process works, see the appendix.)

Ultimately, a satellite image may become part of a geographic information system. As such, it becomes a "layer" of information that can be superimposed on other layers such as topographic contours, geopolitical boundaries, place names, property ownership maps, and so forth. When used in this manner, satellite imagery becomes a powerful tool for land use management and planning (fig. 1.7).

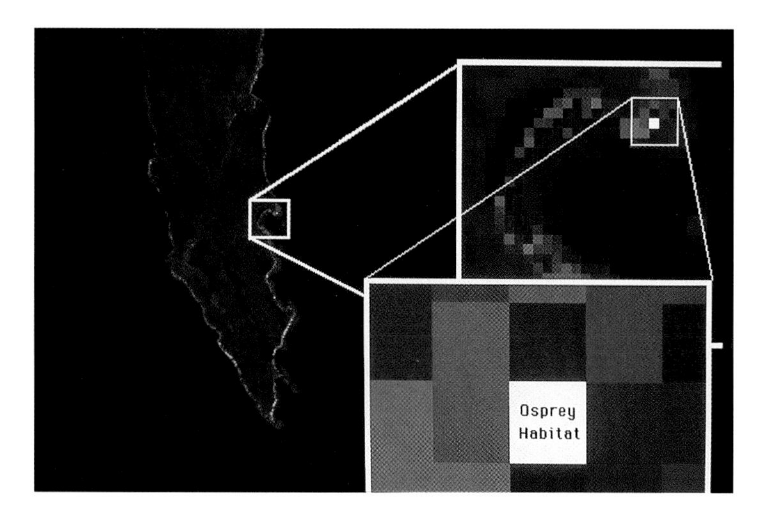

1.6 Osprey Habitat

This false-color image of Petit Manan Point in eastern Maine shows successive close-up views of a satellite image. The satellite cameras "sense" different reflectances from features on the earth's surface, which appear in different shades of red (for land features) and shades of dark (for water features). Part of the value of imaging technology is that it allows a viewer to see the big picture and then zoom in on features of interest. Here the single white pixel shows a 20-meter-by-20-meter ledge area near a brackish pond behind a beach. Ospreys favor such areas for nesting and roosting habitat. *Credit: R. Podolsky, Island Institute*

Many images appear in this book. The detail they contain will reveal whether they were acquired by a camera or a satellite sensor, and for the latter whether the satellite was a land-oriented satellite such as Landsat or SPOT, or an oceanographic or meteorological satellite such as a NOAA Polar Orbiter. In the land-oriented satellite images, you are able to see features as narrow as a country road or a mountain stream. The oceanographic images reveal large-scale patterns in the Gulf of Maine, properties such as sea surface temperature and phytoplankton pigment concentration (see fig. 1.4). Generally, you will see no geopolitical boundaries, but instead you will see the patterns left by nature and by humans for thousands of years—patterns of glacial ice flows, of winds and tides, rivers and rainfall, and the intricate patterns that humans weave into the landscape.

For the past 20 to 30 years, orbiting satellites have been collecting images of the earth, offering us opportunities to visualize the region's interconnectedness and also to appreciate its finite limits. Now, thanks to the explosion of computer technology, its speed and miniaturization, it is possible to view complex images on inexpensive computers in forms simple enough that schoolchildren can use them.

In 1988, the Island Institute in Rockland, Maine, initiated a project to develop its own in-house software system to analyze and manage a library of digital images—satellite images and aerial photographs—of Maine's 4,600 islands and associated coastline. The system is designed for a Macintosh desktop computer with color monitor because of its ease of use and affordability. The software is called GAIA, an acronym for geographic access image and analysis. Many of the thematic maps seen in this book were produced using GAIA.

Two years later the Bigelow Laboratory for Ocean Sciences in West Boothbay Harbor, Maine, in cooperation with the Island Institute, initiated a project called Gaia Crossroads to explore the value of satellite images and hands-on image analysis techniques as a tool for teaching environmental sciences at the

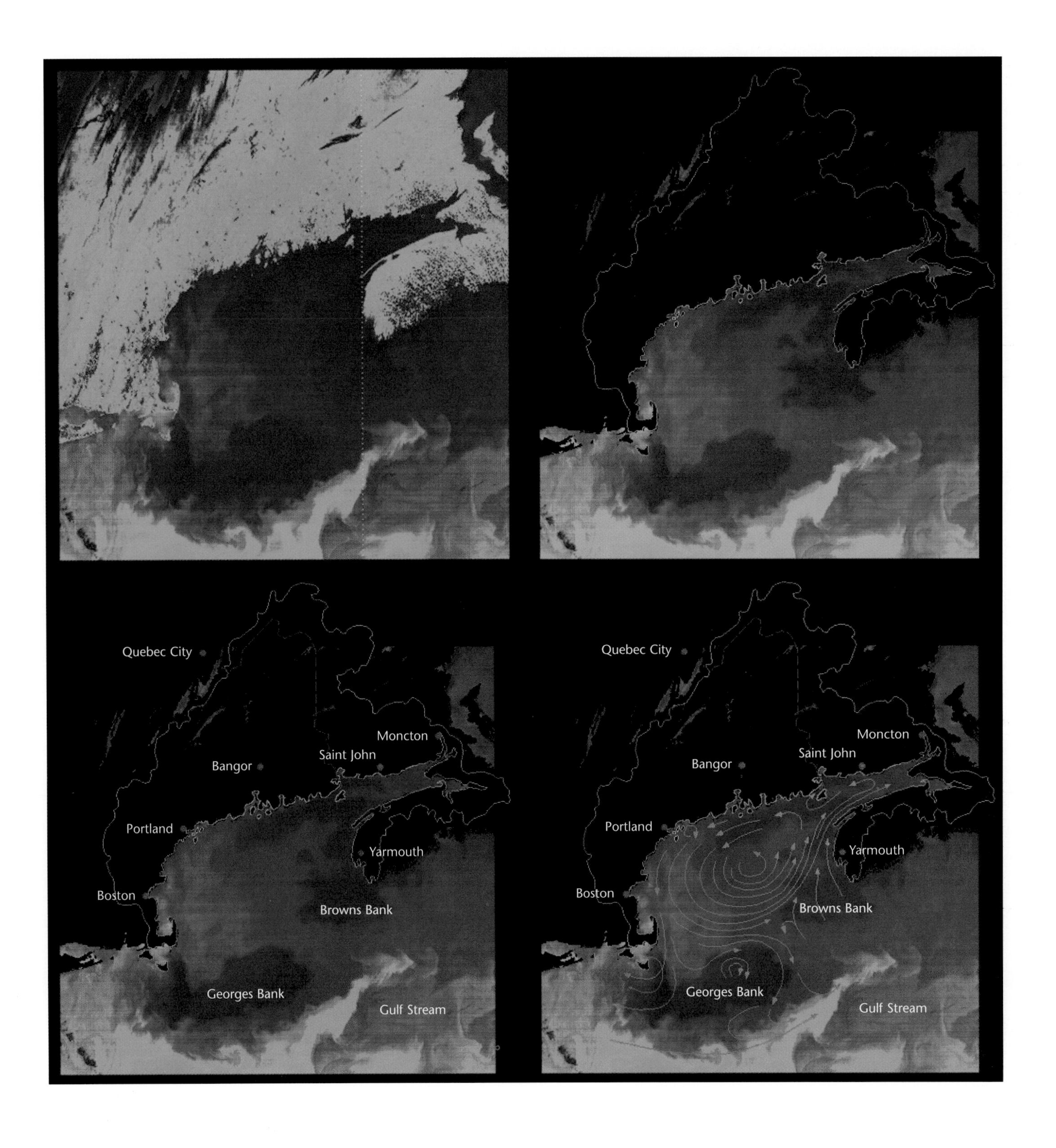

Quebec City
Moncton
Bangor
Saint John
Portland
Yarmouth
Boston
Browns Bank
Georges Bank
Gulf Stream
Quebec City
Moncton
Bangor
Saint John
Portland
Yarmouth
Boston
Browns Bank
Georges Bank
Gulf Stream

1.7 Geographic Information System (GIS) Layers

Geographic information systems are a means of organizing and displaying geographic information using computers to show relationships between different layers of information. In this example an image of the land and waters of the Gulf of Maine is used as base layer (upper left), then the land is simplified and darkened to enhance the view of the water and the terrestrial watershed boundary is added as a vector file that is "registered" on the image data (upper right). Next the major cities and towns and the international boundary are added (lower left), and finally the predominant marine currents are indicated (lower right). *Credit: image enhancement, T. Ongaro, Island Institute*

1.8 Students in the Field

Gaia Crossroads students from Islesboro measure out a 20-meter "pixel" on the ground in a field (above) and then ground-truth a complex littoral habitat of marine intertidal, fringing salt marsh, and upland forest (below). *Credit: Gaia Crossroads Project, Bigelow Laboratory for Ocean Sciences*

elementary, middle, and high school levels. Computers donated by Apple Computer were placed in 20 Maine schools. The computers were equipped with satellite images and the GAIA image analysis software. In the years since, we have been overwhelmed by the response of teachers and students, many of whom had never used computers before, and certainly not for a subject that had been restricted to courses taught in graduate departments at a few universities. This approach to teaching environmental sciences has now been expanded to approximately 100 classrooms throughout New England (fig. 1.8).

Today students in rural Maine towns and elsewhere across the globe are able to operate satellite receiving stations that capture direct broadcasts from weather satellites and display a vast library of images on computer screens. Students at all levels and all ages can study detailed images of the land around their schools, identifying wetlands, forests, fields, and rocky shores, and can see—plain as the nose on their face—the fundamental connections between land and sea, between our backyards and this island earth (fig. 1.9).

The student projects shown throughout this book reflect the work of the Gaia Crossroads Project. Whether these students become scientists or not, remote sensing will affect their lives. By developing an understanding of image technology, these students are gaining an appreciation of their connectedness to ecosystems on all scales, and they are empowered with the tools that will help them act responsibly as citizens of their expanded communities.

1.9 Ovensmouth Marine Habitat

This theme-mapping project was completed by students of the Boothbay GAIA Club and shows the variety of marine and terrestrial habitats in the upper Sheepscot River estuary in an unusual area known as Ovensmouth. A variety of commercially valuable fish and shellfish spawn and spend parts of their juvenile lives in these marine waters. The theme map includes an area that was recently acquired by the Boothbay Region Land Trust. The Land Trust asked the GAIA Club to map land cover, water dispersion, and other natural and developed areas. Ultimately the Land Trust acquired major portions of the upland around Ovensmouth, protecting important terrestrial and marine habitat.

The GAIA Club, which meets after school every Friday, was formed at the upper elementary/middle school level (grades 5 through 8) in the 1993–1994 school year. The students of the club create and enhance maps for various groups and members in the community. According to their advisor, "This club is really a small, non-profit business and has built an extensive list of requests from the community for map projects. The students have learned to work cooperatively with other students who may differ in age and expertise. In addition, they have worked together using several research tools: applying mathematical analysis of spectral bands, ground truthing, examining aerial photographs, comparing data with published maps, and reflecting upon experience. Their written and verbal communication skills have been reinforced and strengthened through the group's activities." *Credit: J. Farrin, grade 5, M. Kidd, grade 8, Boothbay Region Elementary School; advisor, J. Clay*

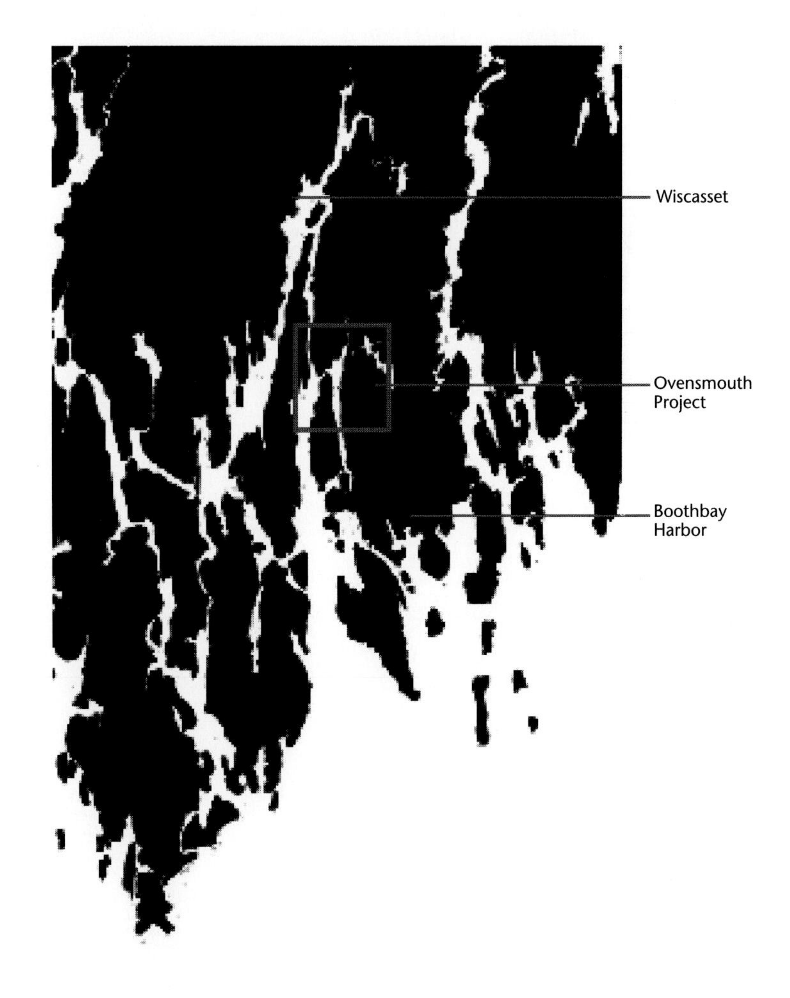

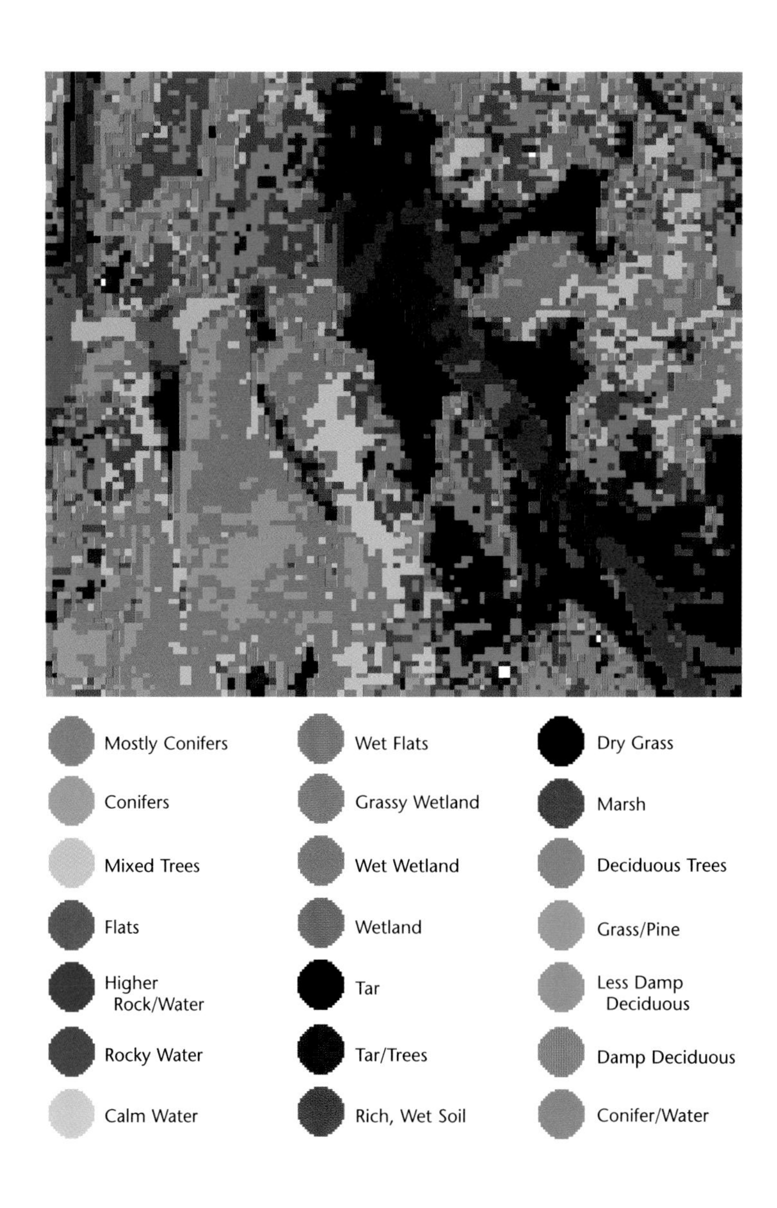
Mostly Conifers
Wet Flats
Dry Grass
Conifers
Grassy Wetland
Marsh
Mixed Trees
Wet Wetland
Deciduous Trees
Flats
Wetland
Grass/Pine
Higher Rock/Water
Tar
Less Damp Deciduous
Rocky Water
Tar/Trees
Damp Deciduous
Calm Water
Rich, Wet Soil
Conifer/Water

2.1 Bedrock Skeleton

Rocks clearly provide the skeleton of the coastline of the entire Gulf of Maine except for the Cape and Islands region. *Credit: P. Ralston*

2

LANDFORMS OF THE GULF OF MAINE

Joseph T. Kelley, Alice R. Kelley, and Spencer Apollonio

Rocks clearly provide the overall skeleton of the entire Gulf of Maine. The shoreline of the Gulf, where rocks are easily visible, exhibits an extraordinary variety of geological features that have captured the imagination of residents and the attention of geologists for more than a century. At the same time, hidden from view far below the Gulf's surface, the geologic diversity of the bottom also shapes submarine habitats. The varying properties of the bedrock framework, imparted during mountain-building events long ago, have determined the present look of such features as the shape of bays and headlands and the number of islands along the rim of the Gulf of Maine and the location of the ledges, shoals, and basins offshore (fig. 2.1).

More recently, during the ice ages of the past two million years, glaciers selectively eroded the rocks of the region and left behind crushed gravel, sand, and mud, which are the materials available today to form our beaches, salt marshes, and countless other coastal and subtidal environments. Occasionally these materials are moved around quickly and dramatically by catastrophic events such as northeasterly gales, hurricanes, or landslides to produce new environments on land as well as on the seafloor. Most of the time these ice-eroded materials are being moved into new locations in slower and more gradual fashion.

Geologic History

Although most of the geologic history of the Gulf of Maine rocks has been relatively uneventful, there have been several episodes of mountain-building activity. Sometime around 430 million years ago crustal movement, analogous to that occurring near Japan or Indonesia today, brought a volcanic island chain into contact with the edge of North America. The collision raised a mountain chain throughout central New England, and the heat associated with the collision formed molten masses of rock that later cooled to form granitic rocks.

This mountainous mass eroded and shed sand, mud, and other sediments into an adjacent sea until renewed crustal movement brought the European-African continents into another collision with North America around 380 million years ago. The layers of sediment were heated into rocks and folded and broken by faults accompanied by earthquakes. A mountain range formed comparable to the Himalayas (which is forming today as a result of the collision of Asia and India), and movement of deep-seated molten rock led to renewed granitic and volcanic rock formation. Most of the rocks surrounding the Gulf of Maine were affected by this period of mountain building. All of the rocks bordering the Gulf are part of this Appalachian Mountain chain, which extends from Georgia to Newfoundland (fig. 2.2). The oldest rocks formed as shallow-water reefs or layers of deep-water sand and mud surrounding volcanic island chains and on the margins of the youthful North American and Euro-African continents. The presence of the volcanic rocks indicates movement of the earth's crust to geologists, although most of the layers of sediment accumulated quietly for hundreds of millions of years with only episodic disturbances.

From 600 to about 350 million years ago the continents of North America, Africa, and Europe slowly converged on the shifting plates of the earth's surface, consuming entirely the original Atlantic Ocean that lay between them. This event, which created the supercontinent geologists call Pangea, left a basement of metamorphic rocks that is the foundation of much of the present Gulf of Maine. Major topographic features of the bottom of the Gulf of Maine, such as Jeffreys Bank, Three Dory Ridge, and Cashes Ledge, were formed by these ancient rocks.

Pangea blocked the escape of heat from the earth's interior and was broken apart by renewed volcanic activity 190 million years ago, as the drift of the earth's surface plates reversed. As large blocks of earth pulled apart, *rift valleys* from the Bay of Fundy formed across the Gulf of Maine and farther south. The rift valleys marked the first stage in the re-formation of the Atlantic Ocean, just as similar valleys in East Africa today mark

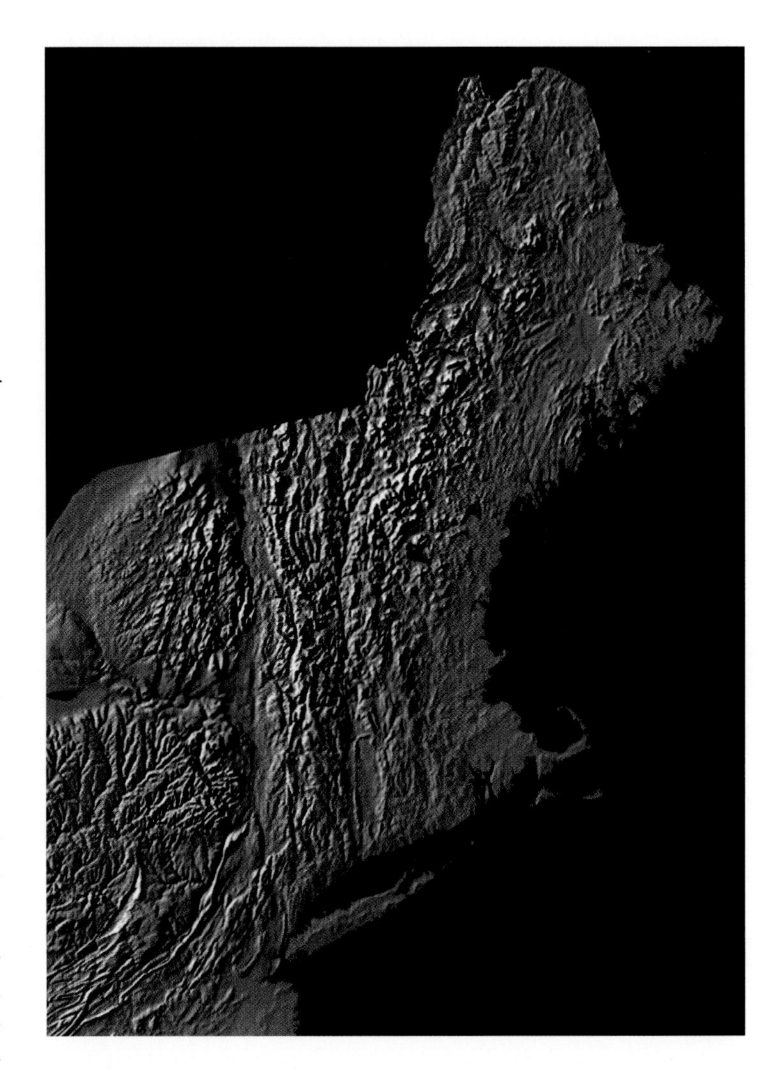

2.2 Appalachian Mountain Chain

The Appalachian Mountain chain extends from Georgia to Newfoundland, a portion of which is illustrated here in an SAR (Synthetic Aperture Radiometer) satellite image. *Credit: U.S. Geological Survey*

the expansion of the Red Sea (fig. 2.3). The rifts were filled by intrusions of *igneous* rocks from the earth's interior. Volcanic activity continued to open the new Atlantic Ocean in a process that continues to the present, increasing the width of the ocean by inches each year. The last granitic rocks formed in New Hampshire and southern Maine around 65 million years ago.

In subsequent epochs, erosion has reduced the height of the Appalachian Mountains greatly. Indeed, the landscape of the Gulf of Maine is one defined by erosive processes acting on rocks that resist erosion differently. Even over small distances, differences in the rate of erosion of various rock types have led to differing heights of the rocks. Granites have generally resisted erosion most strongly and tend to form high points and headlands, while the altered sand and mud layers, the *metamorphic* rocks, are more readily eroded and occupy the bottom of bays and lowlands. The erosion has not occurred evenly throughout the Gulf of Maine, but has been more pronounced in the south than in the north. Thus, we find relatively youthful, unaltered *sedimentary* rocks at the surface in New Brunswick today, while rocks altered by extreme heat and burial (the roots of the ancient mountain chain) are exposed to the south.

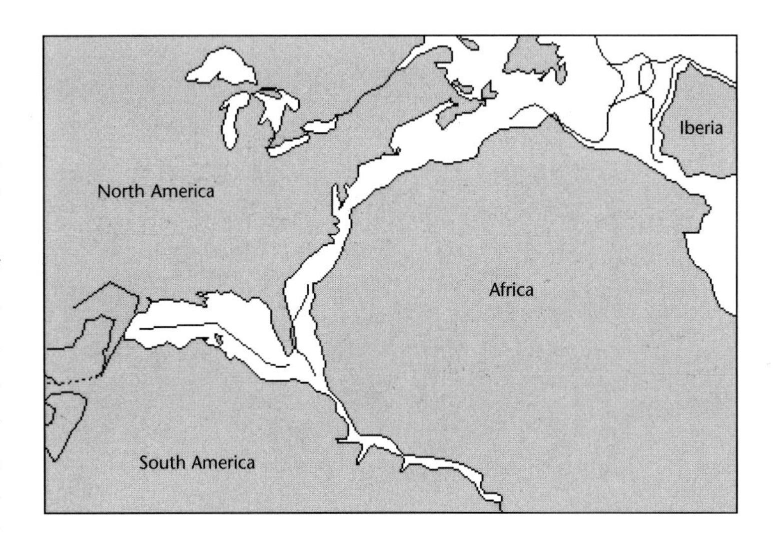

2.3 Pangea

Illustration of the breakup of the supercontinent of Pangea, showing the creation of the Atlantic Ocean and the Gulf of Maine approximately 200 million years ago. *Credit: T. Christensen*

Geological Regions of the Gulf of Maine Shoreline

The landward boundary of the Gulf of Maine stretches from Cape Cod north and eastward along the coast of Maine and then across to and southward along the Nova Scotian shore to Cape Sable. This land-bound edge is incredibly complex; if all its bays, islands, and tidal rivers are traced, it is immensely long, perhaps as much as 7,500 miles (12,000 kilometers). Rocks that form the skeleton or framework of the landscapes of this region determine its largest-scale appearances. Based on the orientation of the rocks and their relative resistance to erosion, we can divide the shoreline of the Gulf of Maine into eight compartments, or regions, that are internally similar but differ from their neighbors (fig. 2.4).

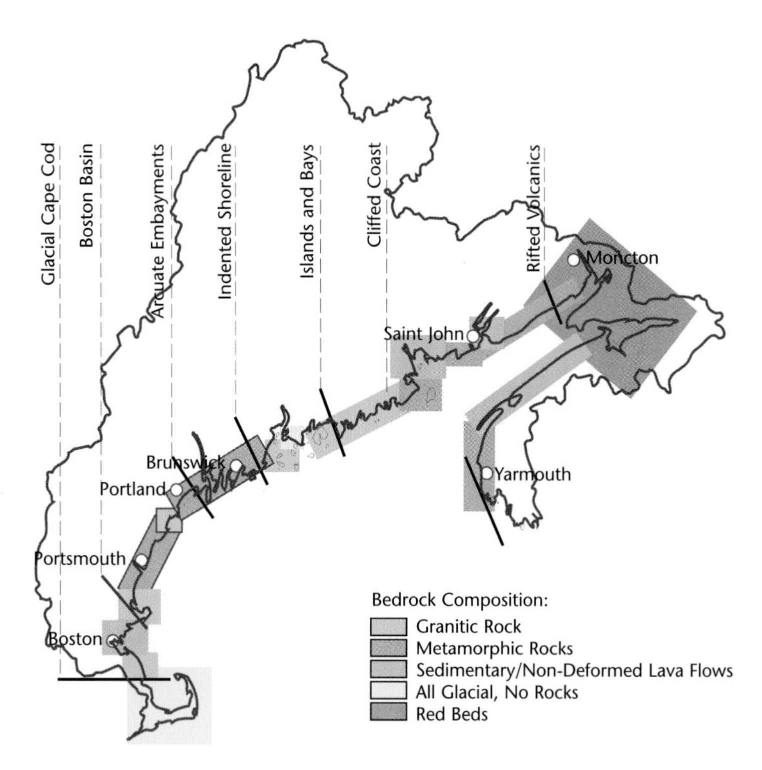

2.4 Geomorphic Compartments

Compartments and major geologic rock types of the shoreline of the Gulf of Maine. *Credit: J. Kelley, S. Meyer*

Rifted Volcanics

Some of the youngest and most distinctive rocks of the Gulf line the southern and eastern shores of the Bay of Fundy. Here, in the Rifted Volcanics compartment, erosion-resistant, black, basaltic volcanic rocks form the highest cliffs in the Gulf of Maine (fig. 2.5). These lava flows welled up from deep within the earth's mantle, and the basaltic rocks are identical to rocks presently creating new ocean crust in the mid-Atlantic Ocean near Iceland. From the ground, individual layers of the flows are visible, as are columns formed during the cooling of the rocks. From space, the straight high cliffs of North Mountain, Digby Neck, and Grand Manan Island are very distinctive features. A fold in the volcanic rocks leads to a bend in the orientation of the coast, and a prominent headland at Cape Blomidon. A most impressive view of the volcanic cliffs is obtained as the ferry from Saint John passes through the cliffs that guard Digby Harbor.

Red Beds

At the head of the Bay of Fundy undeformed red beds, or red sandstones, form weak cliffs that are actively eroding today (fig. 2.6). The eroded sand and mud grains impart the distinctive red hue to the turbid waters of the Bay of Fundy. These red beds formed in rift valleys like the Serengeti Plain of Africa today. Dinosaurs were the kings of the earth when these beds were forming; their fossils are actively collected in the area today. Associated with these rocks are older, though similar, sandstones and coal deposits. Plant fossils associated with the coal represent the early colonization of the land by plants, and are also popular with collectors. The large basins of the Gulf of Maine possess red beds and black volcanic rocks identical to those in the Fundy region. The overall shape of the bottom of the Gulf of Maine is one imparted by the rift basins that form its deeps.

2.5 Grand Manan Cliffs

Black volcanic rocks form the western shore of Grand Manan, the highest cliffs in the Gulf of Maine. *Credit: P. Ralston*

2.6 Red Beds

Sedimentary red sandstone units eroded by wave action produce a host of beautiful and unusual rock formations in the upper Bay of Fundy. *Credit: B. Schwab*

Cliffed Coast

The western shore of the Bay of Fundy and the easternmost portion of the United States are collectively the Cliffed Coast compartment. The rocks here are a complex mosaic of very old metamorphic rocks (among the oldest bordering the Gulf of Maine) juxtaposed with occasional red beds, rift volcanics, and still older volcanic island rocks. The indentations in the coast generally coincide with the occurrence of easily eroded red beds and other sedimentary rocks, while the more ancient metamorphic and volcanic rocks form straight coasts. The mix of rock types in this area, such as in Fundy National Park, results in part from the numerous faults, or locations where earthquakes have ripped up the rocks and moved them. It is important to realize that earthquakes still occur in the Gulf of Maine region, and the Oak Bay Fault, which forms the border between Maine and New Brunswick, is associated with relatively frequent swarms of quakes, some of which have historically been quite damaging to Eastport and Machias.

Islands and Bays of Maine

The straight Cliffed Coast ends abruptly at Machias Bay, from which the Islands and Bays complex extends south to Penobscot Bay (fig. 2.7). The rocks in this compartment vary from relatively easily eroded metamorphic rocks, which underlie the bays, to granitic rocks, which support the islands. The granites were bubbles of molten rock deep within the earth during the collision between Europe and North America, and the rounded shape of Mount Desert and most of the other islands of the region reflect that origin (fig. 2.8). Penobscot Bay is a prominent feature of the compartment, and is the largest embayment in the Gulf of Maine still associated with an important river. (As discussed below, most of the rivers of the region were moved out of their valleys by Ice Age events.) The straight body of water separating Deer Isle from the mainland, Eggemoggin Reach, represents another fault zone, like Oak Bay, where earthquakes have broken up the rocks and led to their removal by erosive processes.

The Indented Shoreline of Midcoast Maine

South of Penobscot Bay, the Indented Shoreline compartment is probably the most distinctive region of the Gulf of Maine (fig. 2.9). The northeast-southwest orientation of slightly more erosion-resistant beds of metamorphic rocks forms the straight peninsulas and chains of islands and shoals; while somewhat more erodible layers of rock form the floor of the many narrow estuaries separating the peninsulas. The subdued nature of this coastline compared with that to the north is a result of the weaker nature of the rocks. Small bodies of granites occur at some of the headlands in the area, as at Reid Beach State Park, but most of the layered coastal rocks visibly demonstrate the deformation of their complex past.

Arcuate Embayments: Portland to Boston

The Arcuate Embayments compartment reaches from south of Casco Bay to Boston and then continues from the south shore of Boston to Cape Cod. Again, relatively easily eroded metamorphic rocks underlie the embayments, while granites protect the many headlands (fig. 2.10).

2.7 Islands and Bays of Maine

This Landsat scene covers approximately 14,400 square miles of the midcoast and upland regions of Maine. The Islands and Bays complex stretches from Vinalhaven through Mount Desert Island (center) to Great Wass (eastern edge). The parallel ridges of foothills of the Appalachians show up nicely between the Penobscot and Kennebec rivers at the left side of the image. Approximately 6,000,000 acres of terrestrial geologic features are displayed here. *Credit: NASA, January 19, 1983*

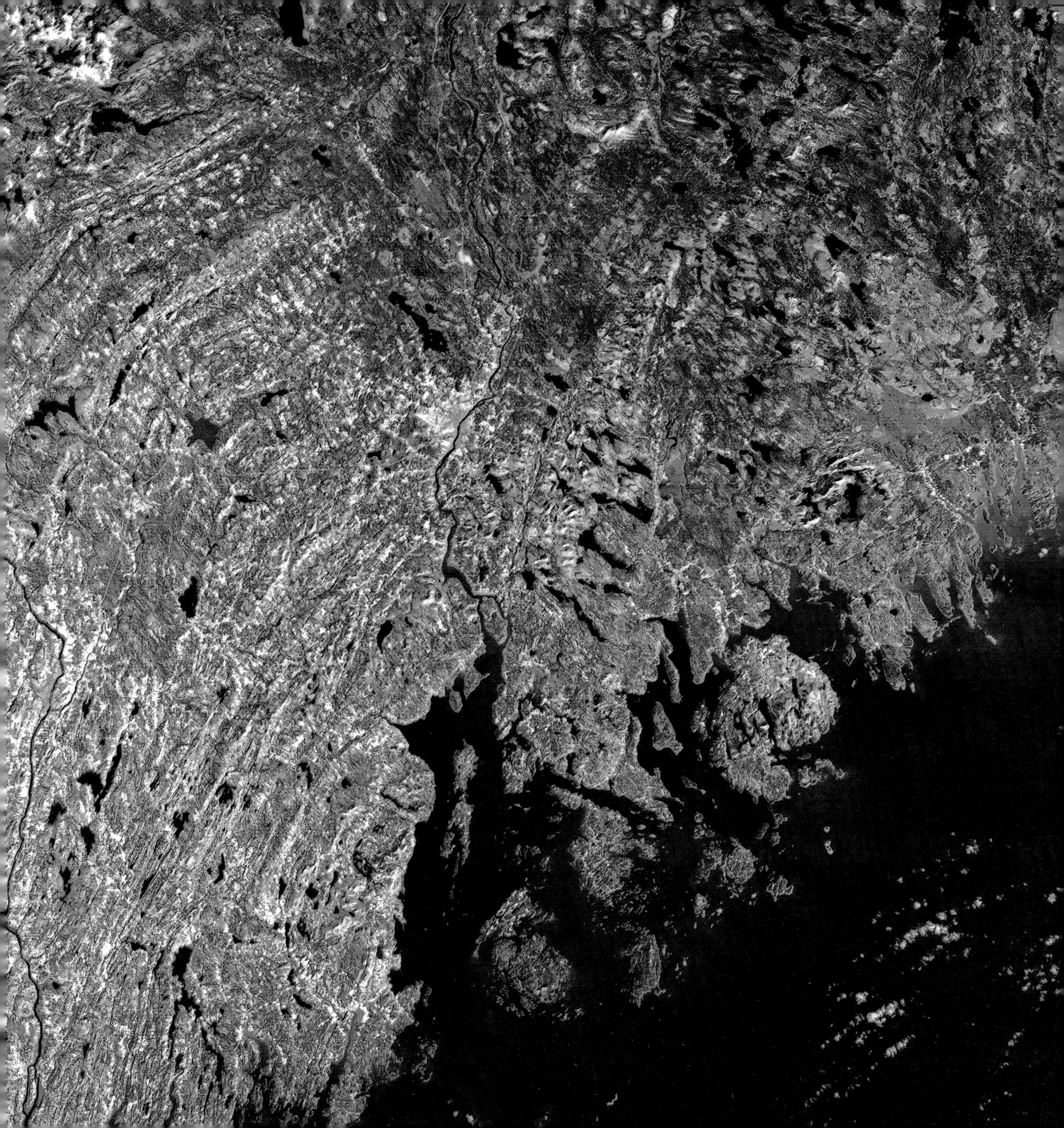

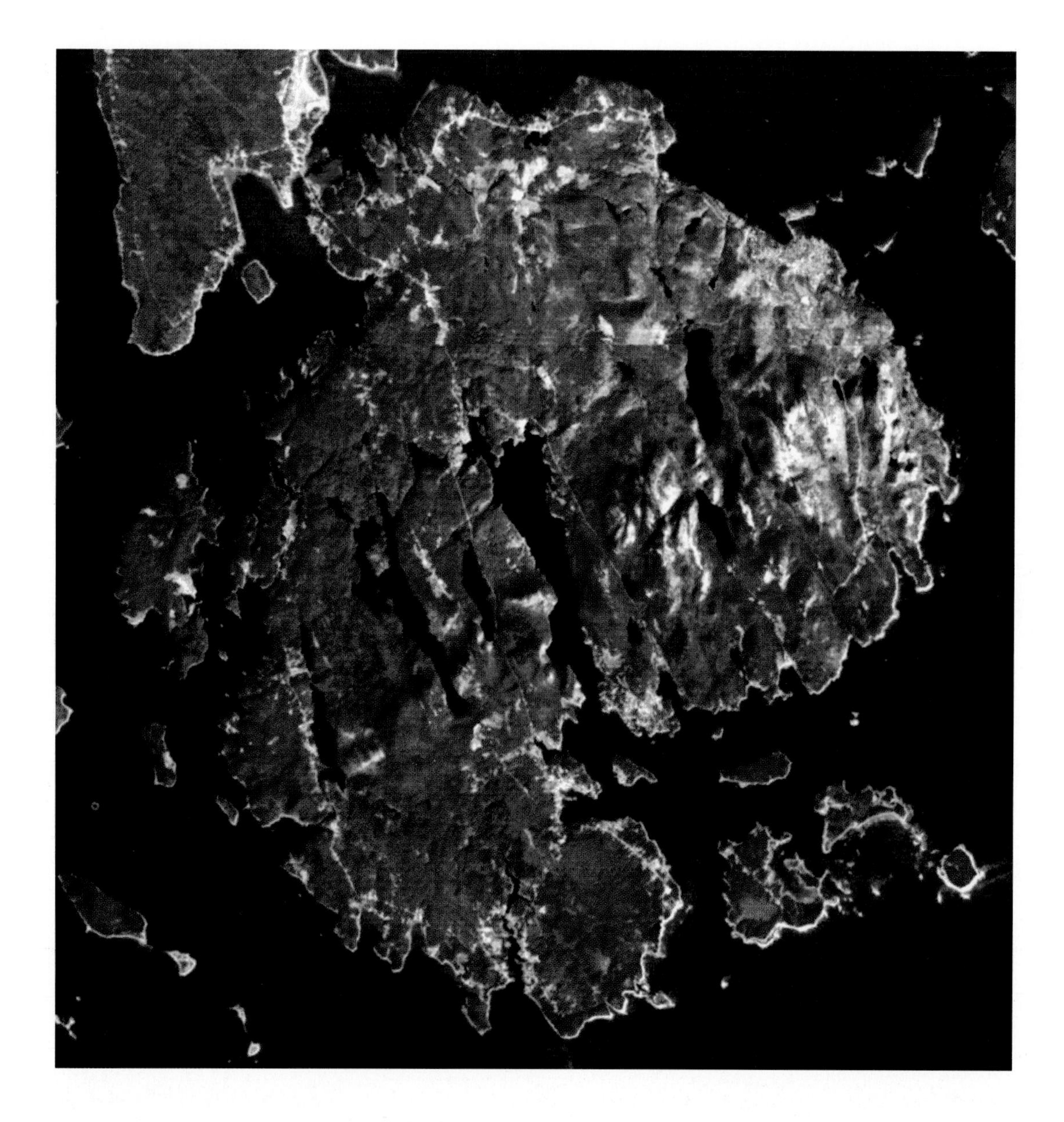

2.8 Mount Desert Island
(left)

False-color image of the rounded granitic shoreline of Mount Desert Island. Individual mountain domes trend from lower left to upper right across the middle of the island. The highest of these mountains is Cadillac, the large white area near the eastern shore. The Cranberry Isles appear in the lower right corner. *Credit: NASA*

2.9 The Indented Shoreline
(right)

This theme map of a SPOT image shows the region from Cape Elizabeth to Boothbay Harbor, including the Androscoggin River entering into Merrymeeting Bay (top center). The paved runways of Brunswick Naval Air Station are visible in red in the center of the image. *Credit: R. Podolsky, P. Conkling, Island Institute*

2.10 Arcuate Bays
(following page)

Landsat image of the coastline of northern Massachusetts and southern Maine from Cape Ann to Cape Elizabeth. The rock type changes at Casco Bay (upper right corner). Lake Winnipesaukee and Sebago Lake emptying into the Presumpscot River are visible. *Credit: NASA*

The youngest rocks in the Gulf of Maine are the granitic rocks that form the distinctive peninsula at Cape Neddick. The metamorphic rocks of the region are composed of relatively monotonous, former deep-water mudstones. Their coast-parallel orientation prevents them from forming peninsulas like similar rocks in the Indented Shoreline compartment.

The Boston Basin

Rocks of the Boston area define their own unique compartment, the Boston Basin. They are mostly easily eroded rocks, hence their low elevation and embayed coast. Fossils from Deer Island, site of Boston's sewage outfall to the Gulf of Maine, show greater affinities with ancient European animals than with North American, and geologists believe that rocks from this area were left behind when the Atlantic Ocean opened and Europe separated from North America. Other rocks in the basin include volcanic materials and a spectacular conglomerate, or rock composed of large boulders. All these rocks are isolated from the rest of the region by faults that surround the Boston Basin.

Shifting Sands of the Cape and Islands

Lacking bedrock of any sort, the southernmost coastal compartment within the Gulf of Maine, Cape Cod, is comprised of Ice Age sand and gravel. The islands to its south, Block Island, Martha's Vineyard, Nantucket, and Long Island, are mantled by Ice Age deposits of varying thickness, but are ultimately supported by loosely cemented material deposited during the past 65 million years. These deposits crop out at Gay Head on Martha's Vineyard and on the eastern portion of Block Island. The unconsolidated, or poorly cemented, nature of much of Cape Cod and the islands makes them prone to extremely rapid erosion by the sea (fig. 2.11). Because of the unusual glacial nature of this region, which occurs on a smaller scale in other places to the north, its formation and dynamics are treated in greater detail later in this chapter.

The Underwater Topography of the Gulf

The United States Geological Survey has called the marine portion of the Gulf of Maine watershed "one of the most striking topographic features off the east coast of the United States," and indeed it is (fig. 2.12). The Gulf of Maine is a rectangular depression of 36,000 square miles (90,700 square kilometers), having an average depth of 490 feet (150 meters). Within the Gulf are 21 deep basins, with Georges Basin reaching a maximum depth for the entire Gulf of 1,236 feet (377 meters). Northeast Channel, a deep trough running between Georges and Browns banks, is 38 miles (61 kilometers) long and 22 miles (35 kilometers) wide, and forms the only deep-water connection into the Gulf of Maine. Georges Bank itself is less than 200 feet (60 meters) deep over much of its extent. In places, such as Cultivator Shoals or Georges Shoals, the bank is barely 13 feet (4 meters) deep, and in heavy weather the seas break dangerously over the shoal ground.

The submarine geomorphology and bottom topography of the Gulf reflect both its underlying rock types and its youth. Perhaps 15–20 million years ago, the area that now includes the Gulf of Maine and Georges Bank lay submerged, probably as a shallow, smooth, seaward-sloping shelf, much like the present-day continental shelf off the mid-Atlantic coast. There was none of the complex topography of banks, gullies, ridges, and basins so common in the Gulf today.

In some way yet unknown the area was then exposed as dry land, probably by an uplift of the sea bottom but perhaps by lowering of the sea level. The exposed lowland was then eroded by streams, and Northeast Channel was probably the main water gap of the drainage system. Later streams of this system eroded the inner border of what is now Georges Bank into the steep slope that now faces northward toward the Gulf of Maine. River erosion also carved out stream valleys that were later excavated even more by glacial ice into the major basins of the Gulf, including Jordan, Georges, and Wilkinson basins.

2.11 Cape Cod and the Islands

The narrow sandy fringe of the cape and islands, which are intensively used by tourists, is etched on this image outlining the fragile zone between land and sea. As some beaches disappear, others such as at Race Point at the extreme tip of Cape Cod are being created. *Credit: Space Shots*

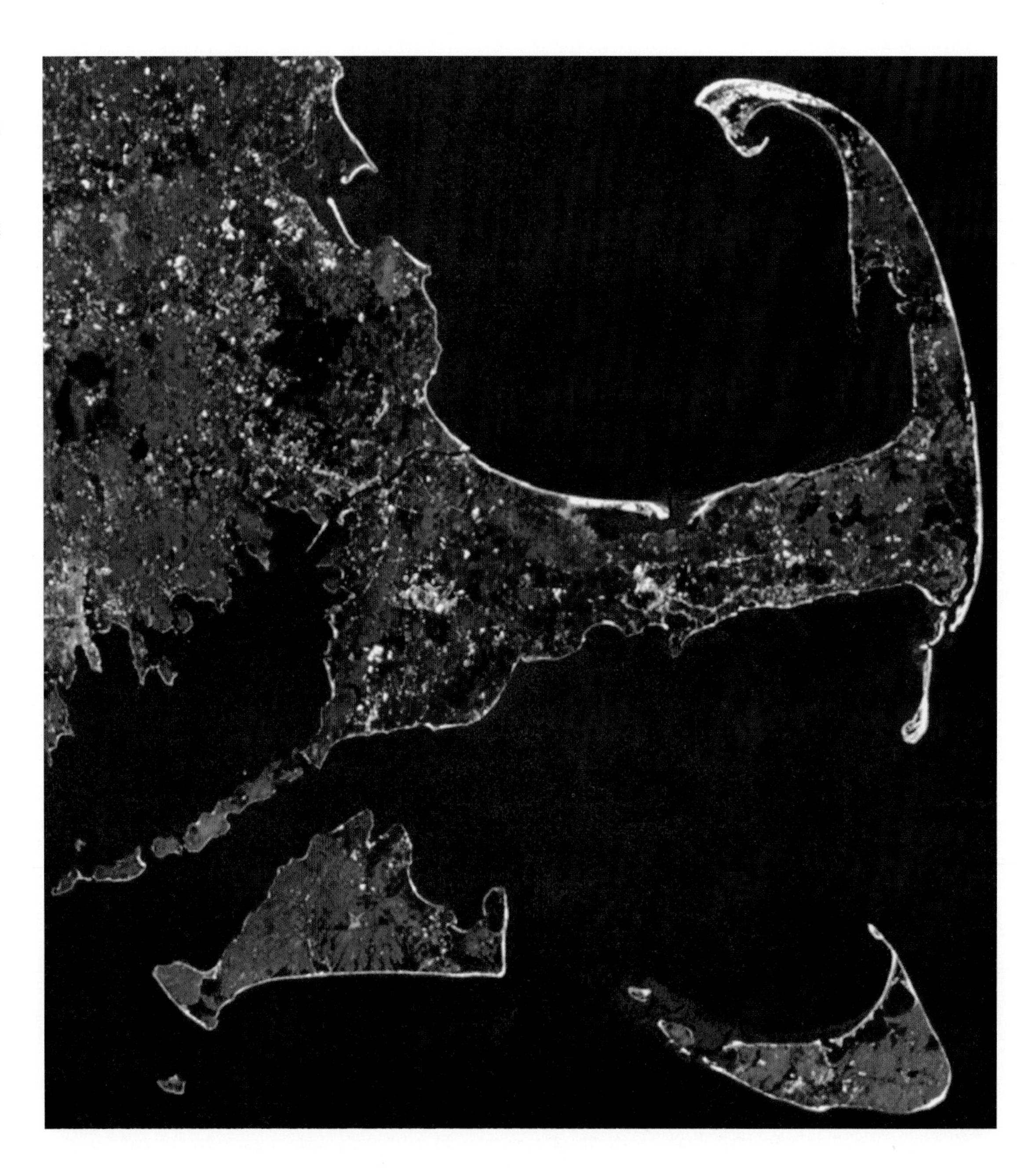

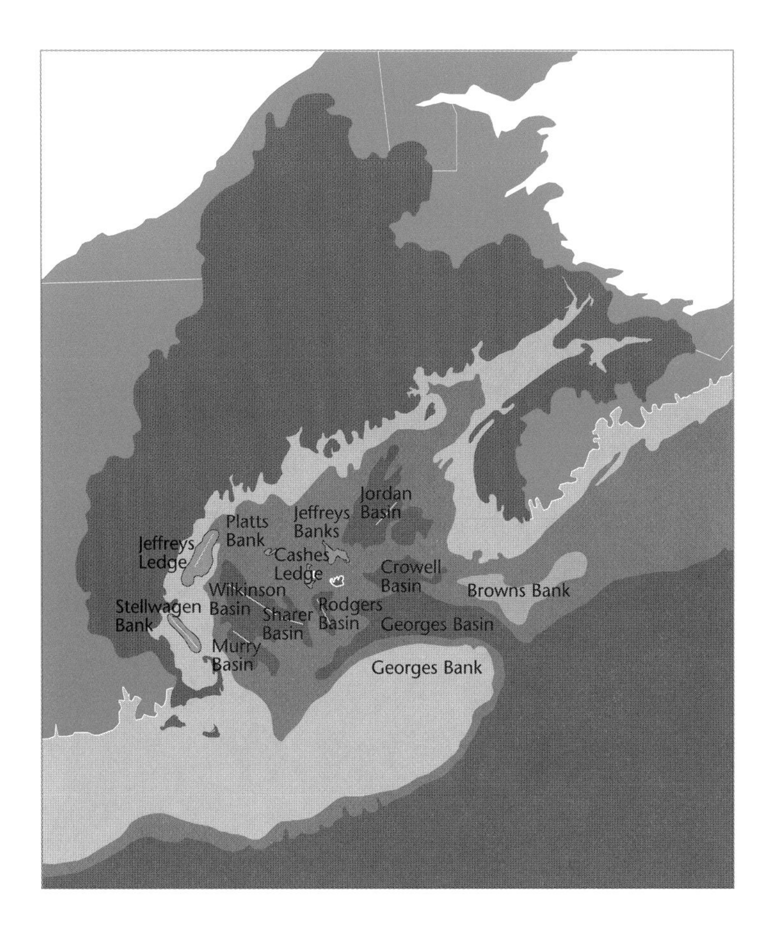

2.12 Underwater Topography of the Gulf of Maine

The bottom topography of the Gulf of Maine is composed of 21 deep basins and innumerable shallow ledges encompassing a great diversity of marine habitats. The major compartments of this underwater region are shown here. *Credit: S. Meyer, Island Institute*

Approximately 13,000 years ago, as glacial ice melted, rising seas flooded into the central part of the Gulf of Maine basin that had been depressed by the weight of the ice and advanced far up the major river valleys to create the DeGeer Sea. New islands were temporarily created along the Maine coast, but Georges and Browns banks were exposed as large capes (fig. 2.13). The land, however, relieved of the great weight of ice, soon began to rebound, and the sea receded. Since then a slowly rising sea has resubmerged older shorelines, but has not pushed as far inland as during the period of the DeGeer Sea.

The great variety of habitats along the coasts and on the bottom of the Gulf is in sharp contrast to the quite uniform shores and bottoms west and south of Cape Cod. Detailed underwater habitat mapping of the Gulf of Maine is a more recent effort than satellite image analysis, but is beginning to provide a wealth of new information about the diversity of these marine habitats (fig. 2.14).

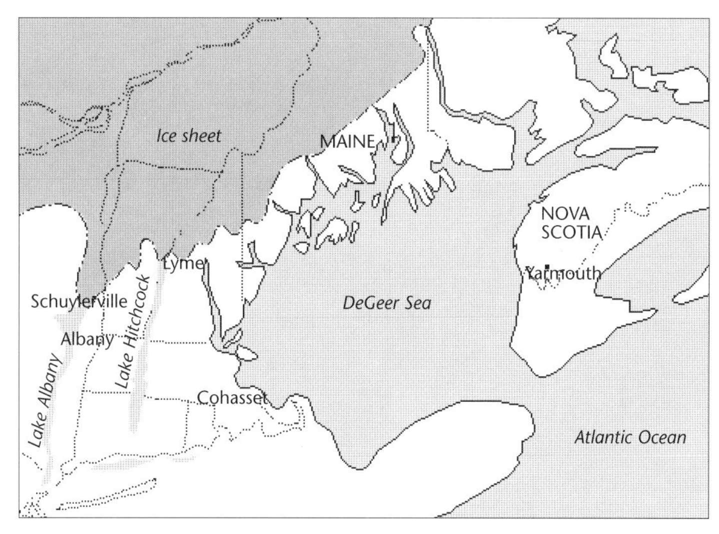

2.13 DeGeer Sea

Although the Gulf of Maine is partially isolated from the Atlantic Ocean, it was even more landlocked 13,000 years ago. Then, in a stage of its development that geologists call the DeGeer Sea, Georges and Browns banks were exposed as dry land, but rising sea level pushed far up the major river valleys such as the Merrimack, Saco, Penobscot and Saint John. Much of eastern Maine was submerged, and Nova Scotia was cut off as an island. *Credit: graphic: T. Christensen; data: J. Kelley*

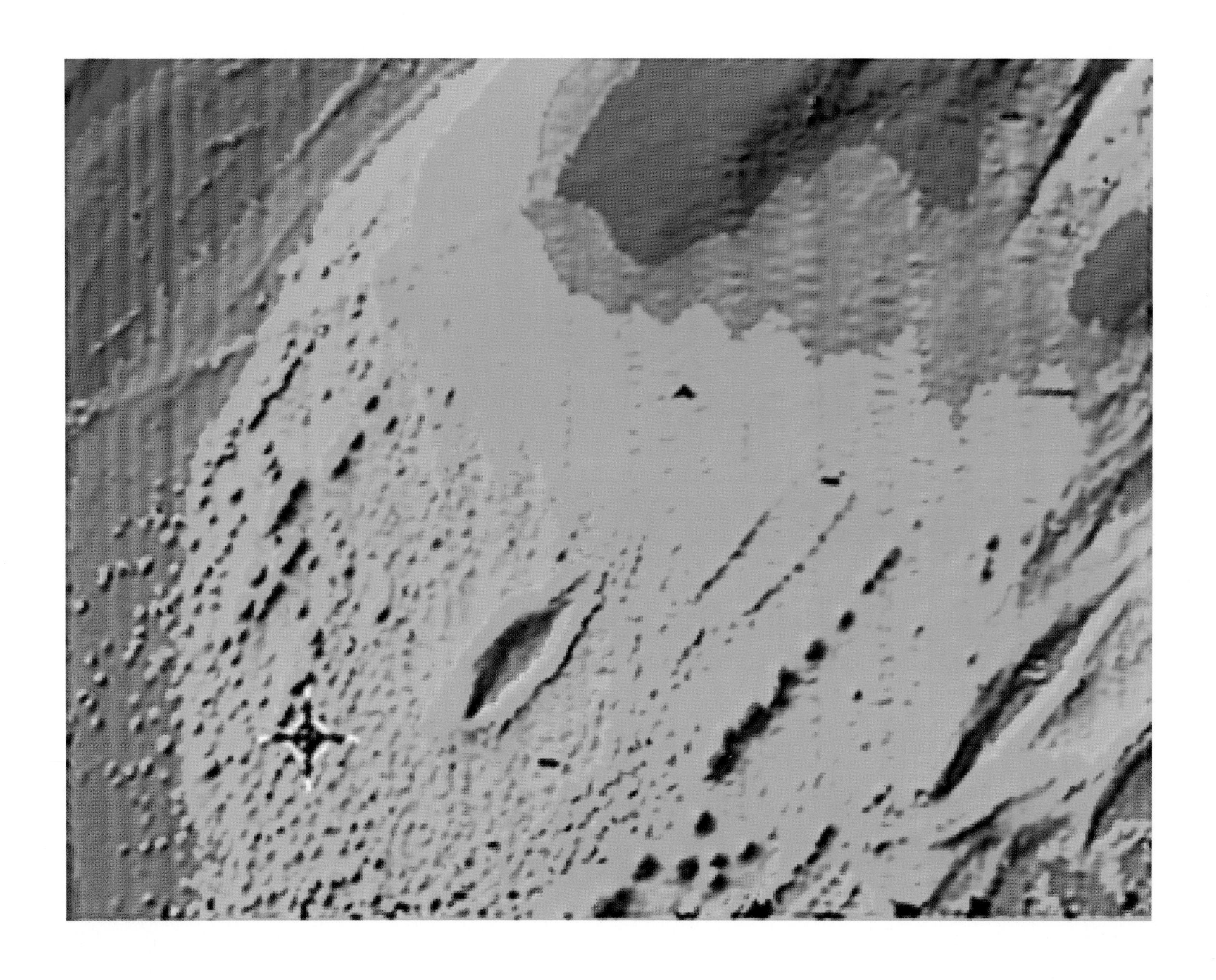

2.14 Seafloor Mapping of Passamaquoddy Bay

Multibeam sensors are used to collect data about the bathymetry (underwater terrain) of the Passamaquoddy seafloor. The Ocean Mapping Group at the University of New Brunswick complements the multibeam data with acoustic data to record the composition (geology) of the seabed. In order to distinguish topography from seafloor composition, a three-dimensional terrain model is constructed and the imagery overlaid on top of it. The pocks in these images are thought to be caused by hydrogen sulfide gas bubbles, which are trapped in the sediments during anaerobic decomposition and periodically erupt out of them. *Credit: University of New Brunswick*

Glacial Processes and Landforms

Earlier we emphasized the importance of erosion in forming the landscapes and subtidal environments of the Gulf of Maine. Hundreds of millions of years of rain and wind and rivers were required to wear down the rocks into their present shape. However, beginning about two million years ago and continuing until only about 11,000 years ago, glaciers influenced the landscape more than any previous event of similar duration.

The passage of glacial ice served to smooth the land's surface by both erosional and depositional processes. Topographic highs were sculpted and rounded by the flow of the debris carried at the base of the glacial ice. Low areas were filled to varying degrees by some of this glacial load. We know from studies of the deep sea that ice ages began about two million years ago. We are not exactly sure why they began, but about every hundred thousand years glaciers have formed in the north and spread south.

Since each glacial advance removes most traces of prior ice ages, it is only in areas removed from glacial scouring, like the deep sea, and at the southern terminus of the glaciation that we find deposits from earlier events. Martha's Vineyard, Nantucket, Block Island, and Long Island represent the end points, or *terminal moraines,* of the most recent ice age as well as of at least one earlier advance. Cape Cod represents a major *recessional moraine,* or location where the ice paused for a significant period during its retreat. Because of its location, the features observed on Cape Cod are primarily those associated with the edge of a glacier. They are depositional landforms created at the margin of the ice sheet that stretched from Hudson Bay to the outer edge of the Gulf of Maine about 20,000 years ago. These deposits are formed of glacial debris shaped by the last movement of the ice front.

The eastward-trending portion of Cape Cod, which is sometimes referred to as the biceps, is called a *glacial moraine.* It is a large mound of till (a heterogeneous accumulation of boulders, gravel, sand, and mud) pushed into place by the leading edge of the ice sheet. Because of the chaotic nature of deposition at the terminus of a glacier, the moraine is as uneven in shape as in the size of the materials that form it. The many small, boulder-littered hills and valleys that define the moraine impart a rough, unfinished character to the Cape. Other small moraines are found throughout New England, but none are as extensive as the Cape Cod moraine.

Immediately south and to the east of the Cape Cod moraine the land is less hilly and is pockmarked with lakes and ponds. As the moraine was pushed into place, small rivers of melting water were clogged with sand and gravel as they flowed away from the ice. The deposition of this material formed an apron of sand and gravel layers in front of the moraine known as an *outwash plain.* The lakes and ponds of this region occupy *kettles* or depressions, formed as the melting water from the east flowed into Cape Cod Bay (which was a lake) and cut distinctive channels across the outer Cape. Ten thousand years of erosion by the sea has left no land where this eastern body of ice once existed, but the productive fishery of Georges Bank owes its sand and gravel veneer to that lost ice sheet.

As we move from Cape Cod along the western margin of the Gulf of Maine, the dominant glacial landforms change from those found exclusively at the edge of a melting glacier to those that are also formed beneath its ice. The many north-south-trending valleys on Mount Desert Island, including Somes Sound, the only fjord on the United States' east coast, were selectively etched by the ice which moved in the same direction. Sebago Lake, west of Portland, was scoured so much that its bottom (greater than 300 feet or 100 meters deep) is below sea level. Although these are the only large-scale glacial-erosional features near the Gulf of Maine visible from space, evidence for erosion by ice at smaller scales is common throughout the region.

In many areas the ground surface is hummocky and boulder-littered due to an irregular carpet of till. Larger depositional features, such as *drumlins,* commonly form hills that are especially prominent around Boston. Here, these large teardrop-shaped mounds comprise the harbor islands (fig. 2.15) as well as some of the city's high points, such as Beacon, Bunker, and Breed's hills. Although drumlins are easy to recognize, the origin of these features is not well understood.

In eastern Maine, extensive blueberry barrens exist on top of what were once sandy river *deltas.* These deltas are found from New Brunswick to New Hampshire, though they are most extensive in eastern Maine. Wherever they are located, they mark the landward extent of the early postglacial sea that formed as the ice sheet retreated across Maine. The sea invaded Maine because the great weight of the ice had lowered the elevation of the land profoundly. Thus, despite the fact that global sea level was lower than today because of all the seawater still remaining in the world's glaciers, all of the lowlands of the state were covered by the sea. Marine mud deposited at this time mantles the surface of Maine's lowlands and extends more than 50 miles (80 kilometers) up some river valleys. Aside from smoothing the landscape, this mud provides some of the only good farmland in the region.

Glaciers not only deposited material on land but also covered the lowlands that would become the Gulf of Maine. The deep basins of the Gulf of Maine and the Bay of Fundy were scoured to their present depth by the weight of a mile of glacial ice. But the ice advance apparently was stopped against the steep north-facing slope of Georges Bank and was deflected eastward to form an immense valley glacier flowing through Northeast Channel and calving icebergs into the deep water over the edge of the continental shelf. Glacial outwash sediments and rock debris were dumped from the southern edge of the ice field onto Georges Bank and Nantucket Shoals.

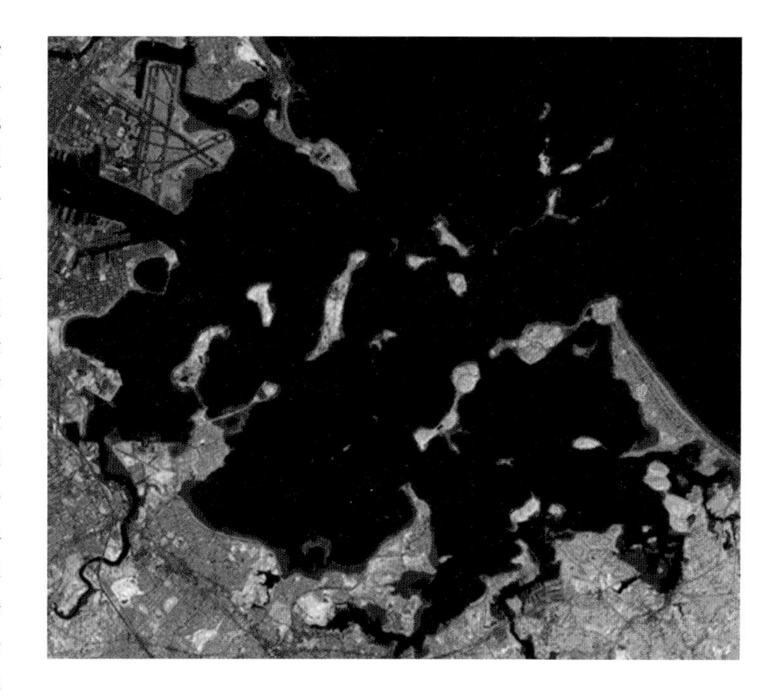

2.15 Boston Harbor Drumlins

The islands of Boston Harbor are an unusual constellation of glacial drumlins surrounded by a rising sea. *Credit: R. Podolsky, Island Institute; GAIA image*

Recent Changes

Following retreat of the ice sheet and withdrawal of the sea, a new landscape began to form. The till, stratified sand and gravel, and marine mud filled topographic lows, often blocking preglacial river valleys. As a result, many rivers were unable to locate their former valleys and were forced over falls and rapids on their way to the sea. This is especially evident near Casco Bay, which was carved by the ancestral Androscoggin and Kennebec rivers. These streams are blocked from entering the sea at Brunswick, Maine, by glacial deposits, and now join in Merrymeeting Bay before passing through rapids at The Chops on a new path to the sea. The Saco, Penobscot, Saint John, and other smaller streams also display this derangement of river drainage by glacial deposits.

Sediments associated with the glaciers of the last 1–2 million years are widespread over the bottom of the Gulf of Maine. They are more than 260 feet (80 meters) thick in Cape Cod Bay, over Georges Basin, and at the entrance to the Bay of Fundy. Elsewhere in the Gulf they are quite thin—less than 20 meters thick. These thin glacial sediments cover most of the bedrock on the bottom of the Gulf (see fig. 2.14).

After the ice melted away, the land rebounded and the sea withdrew to about 200 feet (60 meters) below its present elevation. Many of the coast's earliest archaeological deposits are currently submerged, a testament to the time when sea level was lower and the margin between land and sea extended much further seaward than today. Sea level has risen ever since the land finished its rebound when freed from its weight of ice. The continuing rise in sea level is one of the major and lingering influences of the Ice Age on the shape of the Gulf of Maine today—as we will see in the next chapter.

3.1 Emily, the Tropical Storm

Hurricane Emily as it appeared on August 28, 1993, seen from Channel 1, the visible band of an AVHRR (Advanced Very High Resolution Radiometer) satellite. At the time this picture was taken, off the coast of Georgia, Emily was peaking and dangerous. The leading edge or northwest quadrant of the storm was aimed at Cape Hatteras and its outer barrier beaches. By the time Emily reached the Gulf of Maine, which can be seen at the upper edge of the image, it had been downgraded to a tropical storm, with northeast gale force winds of 50 to 60 knots. *Credit: Satlantic*

3

WAVES, TIDES, AND BEACHES

WEATHER AND CLIMATE INTERACTIONS IN THE GULF OF MAINE

Joseph T. Kelley and Alice R. Kelley

The Gulf of Maine region experiences a highly variable climate, and the changes in weather conditions can be extreme. The land surrounding the Gulf enjoys four distinct seasons, each of which is influenced to a greater or lesser degree by its proximity to this water mass. The ever-changing north-temperate maritime climate is also affected by global weather patterns that can compound even the notoriously harsh regime of the region.

An example of this is found in the legendary winter of 1816. The decade between 1810 and 1820 was already among the coldest on record as the earth was in the grip of the Little Ice Age, a global cold spell that was to last until about the time of the Civil War. Then in 1815 a volcanic eruption at Mt. Tambora, in Indonesia, lifted an enormous quantity of dust into the upper atmosphere. This dust blocked out the sun's warmth and plunged the world into a year of unprecedented cold. For Maine it became known as "the year without a summer," because heavy frosts occurred during each month of the year. Crops failed and people subsisted on the few vegetables they could coax out of the frozen ground. That winter, Penobscot Bay froze solid out to Isle au Haut, Muscongus Bay to Monhegan Island. People rode sleighs to the mainland from the islands, and the temporary ice bridge allowed colonization of islands by fox, rabbits, mice, and other wildlife. The Gulf of Maine region was not the only area influenced by this global phenomenon, but the year goes down in local record books as an example of just how cold our weather can become.

The Gulf of Maine Gyre and the Weather Cycle

The climate of the Gulf of Maine is dominated by the influence of the Nova Scotia Current, a cold offshoot of the south-flowing Labrador Current. The Nova Scotia Current enters the Gulf of Maine between Browns and Georges banks, and its cold water partly counteracts the warming effects of the Gulf Steam. Upon entering the Gulf of Maine, the Nova Scotia Current hooks around to the north and enters the Bay of Fundy before trending south past Maine, New Hampshire, and Massachusetts. The curved arm of Cape Cod finally deflects the current, which in its convoluted path has formed a slow-spinning gyre in the Gulf of Maine and spread its coolness throughout the region. This current and the thermal inertia, or reluctance of the ocean to change its temperature easily, provide the coastal areas with cooler summers and warmer winters than more continental settings to the west.

Summer is the driest, calmest time of the year, with generally weak winds from the southwest or southeast. Such winds are often warm and moist because they circulate over the Gulf Stream current. When warm, moist air crosses the cool Gulf of Maine, it condenses to produce fog. The incidence of summer fog increases from west to east, paralleling the decrease in water temperature. Although summer fogs increase the hazard to navigation in the Gulf of Maine, they also provide an additional source of moisture for coastal plants and animals, and decrease the incidence of fires.

Tropical storms and hurricanes do occasionally occur in the late summer (fig. 3.1), but they are much less significant to the Gulf of Maine region than to areas to the south. That is because the cool water of the Gulf of Maine weakens such storm systems, reducing the scale of winds and waves. Hurricanes also move much more rapidly at higher latitudes than near the tropics, and so spend less time near the coast. In recent years hurricanes Gloria and Bob roared through the Gulf of Maine, but caused relatively little damage except in a few places such as Chatham Harbor on Cape Cod (fig. 3.2). Even the great New England hurricane of 1938 was more of a rain event in the Gulf of Maine than the wind and wave monster it was to coastal Rhode Island and Connecticut. While not the cause of the Gulf of Maine's greatest storms, tropical storms should not be treated lightly, and meteorologists predict that they will become more common in the next two decades than they have been in the past two.

The autumn is a season of change, with the winds adjusting to a predominantly north-northwest direction. The north winds of the fall are strongest just after a cold front passes over the coast from Canada, bringing in polar, continental air masses. Rain usually precedes the passage of the front, and since this meteorological event occurs relatively frequently, the autumn is an important season for rainfall. Next to the spring freshet, rivers entering the Gulf of Maine attain their greatest volume during the fall.

Winter is the most important weather season along the coast. Storms either track up the St. Lawrence River valley and produce sou'westers, or track across the Gulf of Maine and form nor'easters. These storms may last for several days, and the persistent strong winds blowing from a single direction, especially from the east, generate the largest waves that occur in the Gulf. Although the waves often wreak havoc on beaches, many protected bays in the northern portion of the Gulf are shielded from the wind and waves by a cover of nearshore ice from December to April.

Spring is only a milder extension of winter. Storms still occur, though they bring mostly rain even to the northern regions. Snowmelt, coupled with the persistent storms, results in the greatest river contributions to the Gulf of Maine. Between March 31 and April 2, 1987, up to 7 inches (18 centimeters) of rain poured onto Maine where a heavy snowpack had yet to melt (fig. 3.3). The combined rain and snow resulted in the greatest flood observed in more than 100 years on some river systems. Plumes of muddy water could be seen in satellite images of the coast during this time (see fig. 11.9).

The Effect of Weather on Coastal Environments

The principal effect of benign late spring–summer–early fall conditions is to construct or build up coastal environments. The gentle winds produce small waves, which bring sand and shell fragments to beaches and further sculpt fine sediments into

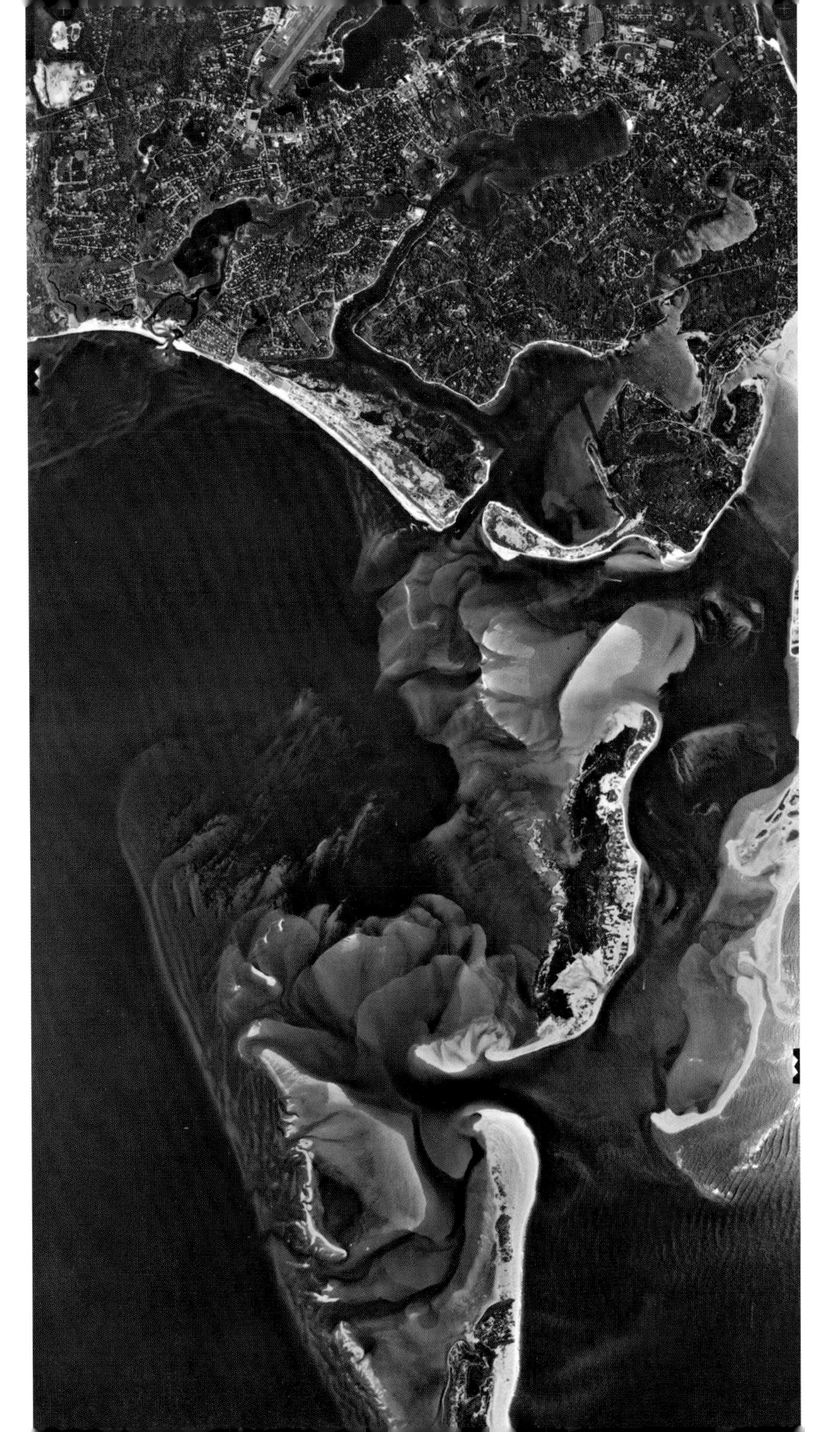

3.2 Chatham Breach

This aerial photo, taken June 5, 1990, shows an example of a storm-caused breach at Chatham Beach, Cape Cod. Reshaping of barrier islands and sand beaches occurs most dramatically during the high winds and wave energy of infrequent storm events. *Credit: James Sewall Co.*

3.3 Saint John River Flood

Due to inordinate flooding on April 3, 1987, the Canadian Centre for Remote Sensing's system was deployed to acquire an SAR (Synthetic Aperture Radiometer) dataset in the Gagetown area toward the mouth of the Saint John River on the Fundy coast. Flooding occurs frequently during the spring runoff and is often associated with river ice jams. During this flood, water levels rose more than 15 feet (5 meters). In this image, open water is the dark tone, forested and agricultural lands are medium gray, ice jams are light gray, and flooded vegetation appears as the very bright features along the shoreline and on the islands. Estimates of the extent of flood damage can be determined by comparing flood images to road maps and property maps, or to other satellite images of non-flood conditions.

In the past, flood delineation was determined through field data collection at given observation points. However, this method is time-consuming and costly during highly dynamic flood events and does not provide adequate areal coverage. With the advent of remote sensing, water resource managers now have at their disposal synoptic coverage of flooded areas using aerial and satellite imaging systems to supplement point source data. Radar remote sensing systems are particularly attractive tools because they allow acquisition of flood data during or near peak flood conditions even under unfavorable weather conditions or darkness. *Credit: D. Werle, AERDE Environmental Research, Halifax, Nova Scotia, Canada; see also "Flood Mapping Using Simulated RADARSAT SAR Imagery" by Robert Leconte and Terry Pultz, Canadian Centre for Remote Sensing, Ottawa, Ontario*

dunes, especially in the southern Gulf of Maine. The spectacular dunes in the Provincelands at Cape Cod National Seashore attest to the importance of the wind in areas with an abundance of sand. Even at smaller beaches, like Plum Island National Seashore, Massachusetts, or Reid Beach State Park, Maine, large dune fields dominate the littoral environments. Although the sand for our coastal dunes clearly comes from the beach, it is instructive to observe that the predominant wind is from the north-northwest, or offshore for many regions. Thus, dune fields are often scalloped on their backside by seaward-oriented, parabolic dunes marking the return of some wind-blown sand to the sea.

The direction and strength of the wind influence more than just the shape of sand dunes; the major effect of the wind is to produce waves. As waves approach shore they encounter islands, which profoundly alter their direction. This change in the orientation of waves by islands and shoals is called refraction, and is most manifest as the encircling necklace of wave crests surrounding islands (figs. 3.4, 3.5). If the islands are near a sandy coast or near the mouth of a river, sand and gravel are moved around the islands, connecting them to the mainland or to one another with a beach called a *tombolo*. Tombolos like the one on the Barred Islands near the mouth of the Penobscot River are unique to rocky coastal regions like the Gulf of Maine (fig. 3.6).

Because of the reorientation of waves by islands, they often approach a beach at an angle. As a result of the momentum of the wave, sand is moved along the beach in the same direction as the wave. This longshore movement of sand results in the formation of a sort of beach known as a spit, attached to land at one end and ending in the ocean at the other. Within the Gulf of Maine, southern Massachusetts and Cape Cod possess the largest spits. Plymouth Harbor is protected by spectacular sandy spits (fig. 3.7), as are Chatham and Wellfleet. All of the fist of Cape Cod west of Pilgrim Springs, including Provincetown and the Provincelands, constitutes a grand spit that has formed during the past several thousand years. At Race Point, a new spit is developing and growing to the west.

Source of Beach Material

Where spits grow and new land is formed, some old land must disappear. The source of sand for the spits of the Gulf of Maine is typically an eroding glacial deposit. Because Cape Cod and its islands are composed exclusively of such material, it has the finest examples of spits (fig. 3.8). To supply its growing beaches, the outer cliffs of the Cape are eroding at a rate of 1 meter (3 feet) per year. The many islands of Boston Harbor are glacial deposits now connected by gravel spits as a result of their long-term erosion by the sea (see fig. 2.15). Nantasket Beach on the south shore of Boston received so much sand from nearby eroding glacial deposits that beaches surrounded and protected several of its drumlins. These drumlins are preserved by their beaches today, with only steep cliffs on the sides of the drumlins to indicate their stormier past.

Though wave erosion of glacial deposits has formed most of the beaches in the Gulf of Maine, some large beach systems off river mouths have had a different origin. Large rivers like the Kennebec, Saint John, Saco, and Merrimack flow through vast watersheds full of sandy glacial deposits. Shortly after the Ice Age ended, when sea level was lower, these streams cut deeply into those deposits on their way to the distant shoreline. There, in present water depths of 200 feet (60 meters), they deposited extensive sandy deltas. As sea level has risen, waves have reworked these deltaic sands into today's magnificent beaches on the western edge of the Gulf of Maine such as Wells, Popham, and Reid Beach state parks, Old Orchard Beach, and Plum Island National Seashore (fig. 3.9). The many islands protecting the Penobscot River mouth prevent waves from forming beaches in Penobscot Bay. An important question facing geologists today is whether these highly dammed rivers still contribute much sand to the coast.

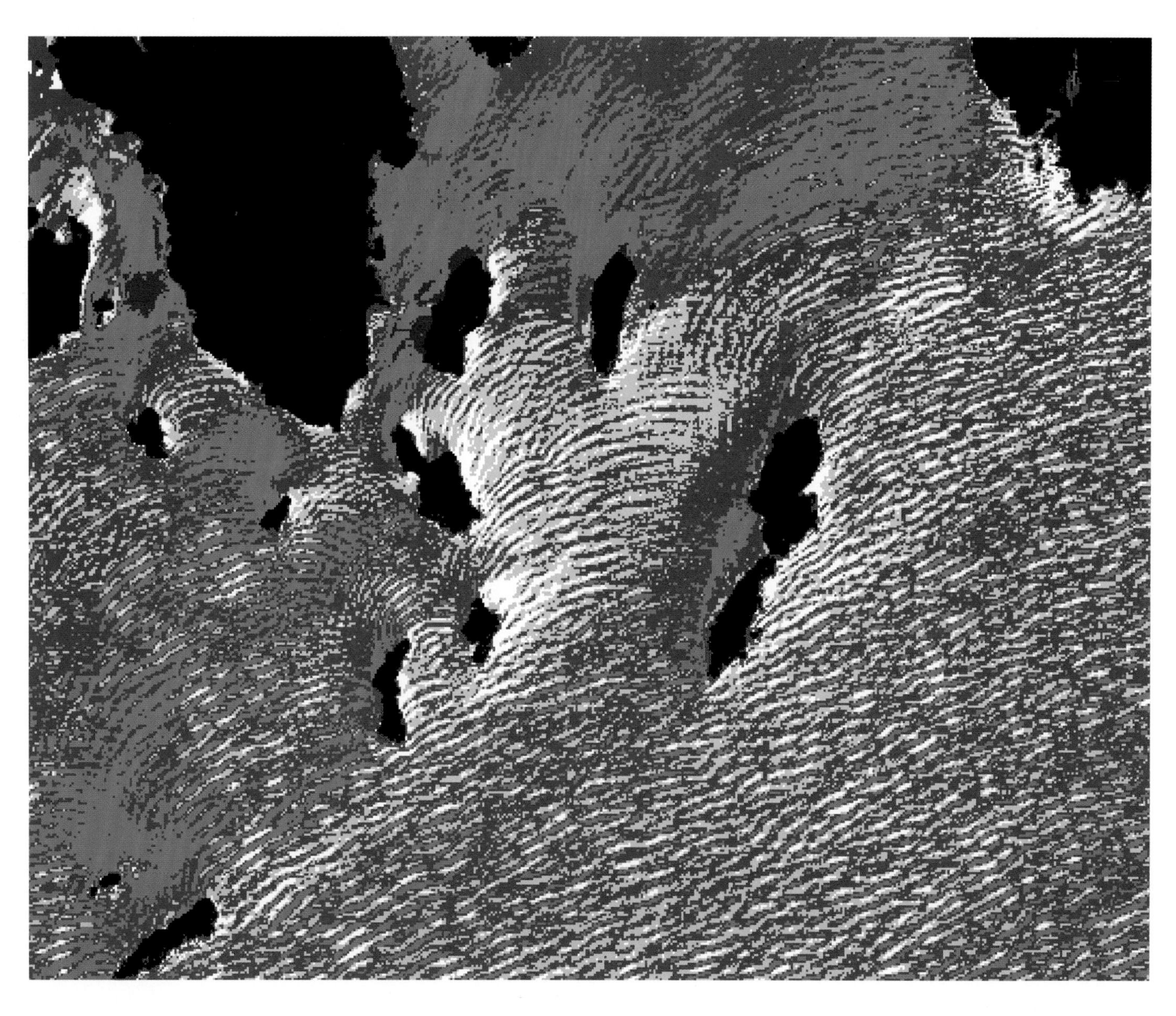

3.4 Wave Refraction

This satellite-enhanced image shows the effect of rocky islands in redirecting the force and energy of onshore waves entering Machias Bay in eastern Maine. *Credit: S. Meyer, Island Institute; GAIA image*

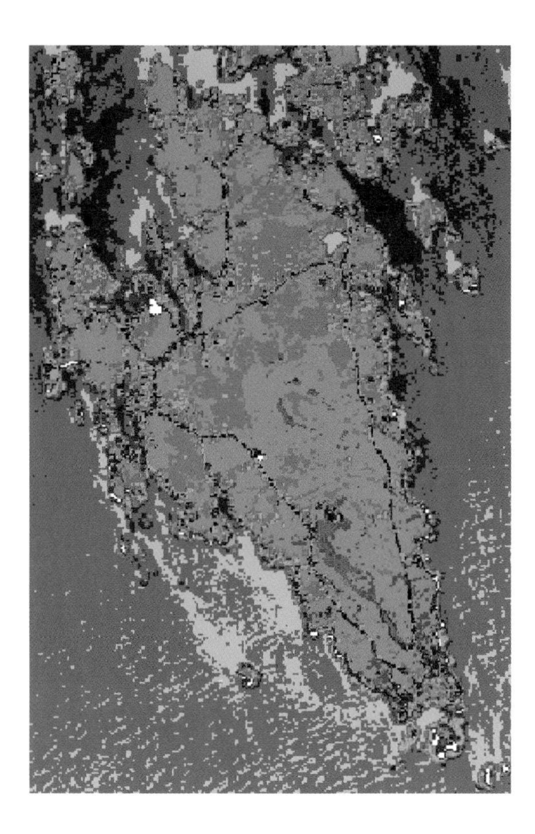

3.5 Wave Climate, Southport Island, Maine

This Gaia student project shows the wave climate around the exposed peninsula of Southport Island near Boothbay Harbor. Areas of quiet water (in dark blue) in the shelter of smaller islands and in coves alternate with high wave energy zones around the outer shores where white wave crests are plainly visible. *Credit: J. Cross, grade 6, Boothbay Region Elementary School; teacher, J. Clay; courtesy of Gaia Crossroads Project, Bigelow Laboratory for Ocean Sciences*

3.6 Tombolo Beach

Doubled-sided beaches such as this one, which connects the Barred Islands in northern Penobscot Bay, are unique to rocky coastlines. *Credit: P. Ralston*

3.7 Plymouth Harbor Spit
(left)

This Landsat image shows the elegant sand spits that have been created by longshore currents, which run parallel to the shore and provide a barrier for the salt marsh and protected subtidal environments of the inner harbor. *Credit: R. Podolsky, Island Institute; GAIA image*

3.8 Nantucket Island Beaches and Shoals
(right)

Glacial sand and gravel deposits form Cape Cod and its islands. These deposits are sorted by waves and tides to create the region's most stunning beaches, rich shellfish beds, and treacherous shoals. *Credit: T. Ongaro, Island Institute; GAIA image*

3.9 Sand Beaches, Southern Maine

Aerial view of Wells Beach, part of the Arcuate Bays region of the Gulf of Maine coast, showing the intensity of development on this landform and its proximity to sensitive salt marshes behind the narrow beach. *Credit: P. Conkling*

Many beaches are not made of sand but are covered with cobbles and boulders. While some of these remain connected to an eroding, coarse-grained glacial deposit, others in more wave-exposed locations may derive their cobbles from bedrock itself. Such beaches are most common on the eastern side of islands, where unimpeded waves are most powerful. Monument Cove Beach, in Acadia National Park, obtains its boulders from granite cliffs driven apart by waves.

Finally, in some sand-starved coves, the only material available to form beaches is broken seashells from animals living on the rocky bottom. Sand Beach in Acadia National Park is composed of 60 to 70 percent shells, mostly belonging to blue mussels, barnacles, and sea urchins. Periwinkles and clam shells may dominate the shell beaches in more protected environments on the western sides of islands (fig. 3.10).

3.10 Shell Beach

This Maine island shell beach is dominated by periwinkle and whelk (marine snail) shells. *Credit: P. Conkling*

Tides, Marshes, and Dams

Tides are perhaps the most striking feature of the Gulf of Maine to the casual visitor. At the south near Cape Cod tides seldom range over four feet (little more than a meter), but they increase constantly as they go eastward. Immediately to the north of the Cape there is a tidal range of seven to ten feet (2–3 meters); in the neighborhood of Passamaquoddy Bay they reach 28 feet (8.5 meters), and as much as 50 feet (15 meters) in the Bay of Fundy—the highest recorded tides in the world.

The main reason for these exceptional tides is that the Gulf of Maine system, out to the edge of the continental shelf, has a natural period of oscillation of a little over 13 hours, and so is nearly in resonance with the 12.4-hour period of the moon-driven tide surging in from the North Atlantic. The range of tide in the Gulf of Maine is also a function of a number of geographical features of the Gulf. The shorelines converging toward the head of the Gulf is one; the oscillations created by the Bay of Fundy is another. As the rising tidal waters enter the Bay of Fundy, its unique characteristics—its length, depth, and shape—act to amplify the tidal range down the entire shoreline of Maine and along Cape Cod's northern shores.

Tides in the Gulf of Maine are semidiurnal, meaning that highs and lows occur approximately twice a day. Because of their familiarity we tend to think of our semidiurnal tides as typical of all tides, but in many parts of the world, such as in the Pacific region, tides may occur only once a day. Canadian *Sailing Directions* for Nova Scotia and the Bay of Fundy note that when the tides surge into the Gulf of Maine from the Atlantic Ocean they undergo considerable changes. Along the coast from Boston to Bar Harbor, for example, high water occurs nearly four hours later than on the Atlantic coast of Nova Scotia, and the range of the tide is nearly double. The rapid changes in the timing and ranges of high water are very apparent around the south end of Nova Scotia.

Tides are an important process that builds up our coastline. Tidal energy does not add much to our beaches; rather its influence is spent on the mudflats and salt marshes. In parts of the coast sheltered from large waves by beaches, islands, or shoals, the periodic movement of the tides brings mud into coastal embayments just as waves often contribute sand. Very often, muddy glacial deposits eroding along the coast are an important source of mud to our clam flats and marshes (fig. 3.11). Where tides are great, great sandy bar systems often exist near the low-water line. The orientation of the bars is indicative of the onshore-offshore direction of water movement. The turbid red waters of the Bay of Fundy region are produced by the constant scouring of the local red rocks by tidal action. Locally, when the tides change in bays with islands blocking their channels, like Cobscook Bay and Passamaquoddy Bay, whirlpools form to facilitate the changing water levels.

It is the enclosed nature of the Gulf of Maine, coupled with the irregular shape of its coastline, that permits the moon's gravitational attraction to create our exceptional tides. That is why construction of a proposed power-generating tidal dam across the Bay of Fundy, by changing the shape of the Gulf of Maine, would profoundly alter the tidal regime. It has been suggested that a change in tidal amplitude—such as could occur from the proposed tidal power development—would cause a number of significant changes in the Gulf: in the speed of coastal currents, incidence of fog, patterns of primary productivity, flushing rates in coastal estuaries, distributions and settling patterns of larvae of shellfish and finfish, and the locations and magnitude of ocean fronts. Such changes have undoubtedly occurred as a result of the increase of tidal amplitude since 10,000 years ago, and to alter the situation suddenly and dramatically would almost certainly have major unforeseen consequences for the entire Gulf of Maine ecosystem.

The Importance of Annual Storms

The benign winds, waves, tides, and currents that operate most of the time are constructive processes that build up our coastal environments. Mud is added to our tidal flats and marshes from eroding bluffs of Ice Age sediment; spits and other beaches are lengthened by waves and currents. In the late fall and winter, however, storm systems are more powerful than in the summer, and the coastal environments must adjust to new conditions.

The large, closely spaced waves of a northeast storm do not bring sand to add to the beaches. Instead, the colossal waves, raised even higher by the low atmospheric pressure of the storm, reach farther up onto the beach, saturate it with water, and often remove sand. Even the sand dunes may be assaulted by the waves of a winter storm at high tide. During the most powerful events of any given year, the smaller dunes are breached and storm waves roll unimpeded across an entire beach. They may remove sand from the seaward side of a beach and deposit it on top of the salt marsh on the landward margin. When the storm is over, the beach may appear wider and lower, and when closely examined by a geologist, it will likely be found to have moved somewhat landward. These occasional storm events similarly have an exceptional impact on eroding bluffs, causing them to retreat

3.11 South Lubec Mudflats

This view of Lubec Flats, north of Quoddy Head in Cobscook Bay, shows a major area of quiet waters where fine-grain sediments are deposited by vigorous tidal action. Here in the summer and early fall shorebirds congregate in enormous flocks to gorge on marine life before crossing the Gulf of Maine. *Credit: C. Ayres*

3.12 Groundhog Day Storm

The northeast gale of February 2, 1978, which came to be known as the Groundhog Day Storm, will be remembered for its destructive fury. It peaked near noon on a spring high tide, and its sustained winds, which exceeded 70 knots offshore, built up tremendous waves that battered the outer islands. This picture shows 40–50-foot (12–15-meter) waves breaking over the tops of trees on the southern end of Hurricane Island in midcoast Maine. *Credit: N. Elam*

and yield most of their annual contribution of sand and mud to the littoral zone.

The greatest storm to reach the Gulf of Maine in the past century occurred in February of 1978. It reached its peak on a spring high tide, when the gravitational pull of the moon was exceptionally strong (fig. 3.12). The resulting water level was more than 10 feet (3 meters) higher than normal. Property damage all along the shoreline was extreme where houses had not been built to withstand such a storm. Following this event, Maine enacted legislation to prevent construction of houses in areas that could be reached by storms with a predicted recurrence interval of once in a century.

Storm events thus move our coastal environments landward, toward higher ground. Beaches are not destroyed by this movement, nor are marshes and flats. Instead they are shifted, reinforced by their adjustment to the rigors of winter's storms. It is this adjustment that has permitted them to endure the final modern process, the long-term rise of sea level.

Zonations within Embayments and Sea Level Rise

Even to the casual observer, it is obvious that the modern coastal environments are arrayed in a highly regular fashion in any embayment along the coast. The outer area, where waves are most powerful, is often rocky and long since swept clean of glacial deposits. Only boulder beaches and bedrock can survive the rigors of such wave exposure. Even the seafloor is bare rock in open coastal locations. If some protection exists from waves, as at Saco Bay, Maine, or if an abundance of glacial sand is present, as at Cape Cod, beaches may be present, but these too are high-energy wave environments. Thus geologists consider the outer portion of an embayment a wave-dominated zone (fig. 3.13).

In the most landward and protected regions at the head of an embayment, salt marshes and tidal flats flourish. Here waves are less important, and tides and rivers build up the muddy elements of the coast. Geologists refer to this region as the tide- or river-dominated zone of an embayment, and it is the area of most extensive salt marshes and mud flats. Glacial deposits in this area are generally protected from erosion by the sheltering effect of the marshes and flats.

Between the wave- and tide-dominated zones lies a mixed-energy zone, which may occupy the largest portion of an estuary. This is the region where glacial deposits are actively eroding and supplying new material to the coast. Marshes and flats, where they exist in this zone, are usually quite small and almost always eroding.

The link between these zones is the rising level of the sea. The wave zone has been exposed for the longest time to the rigors of the sea and all glacial deposits are gone. Beneath the beaches, and occasionally on their seaward sides, geologists have long observed deposits of salt marsh peat and tidal flat mud, indicating the landward shift of the coastline. Environments that once flourished at the head of a bay are now exposed beneath the environments at the bay mouth. The rising level of the sea is moving all the coastal environments landward. In time the quiet environments of the tide-dominated zone will become more exposed by rising sea level and waves will begin to erode the marshes and flats. As their protective influence disappears, the glacial deposits will themselves become exposed and erode and provide material for new tidal environments to the west. Eventually the rising sea will have swept the glacial deposits away, and only sand and boulder beaches will survive. All the coastal environments will be drowned and become a productive seafloor, while the present upland will have evolved into a new coastline.

People and the Future Shoreline

All the elements of our present coastline are dynamic, changing features. All are "new" and have evolved into their present condition as a result of the rising level of the sea in the Gulf of Maine. Everything we take for granted and enjoy along our coast, the recreational beaches, productive marshes, tidal flats and nearshore waters, even the scenic vistas are changing artifacts of the present elevation of the sea. Native Americans from several thousand years ago would not even recognize many parts of the present coast, so great have been the changes. Even the brief period during which we have made maps and recorded the

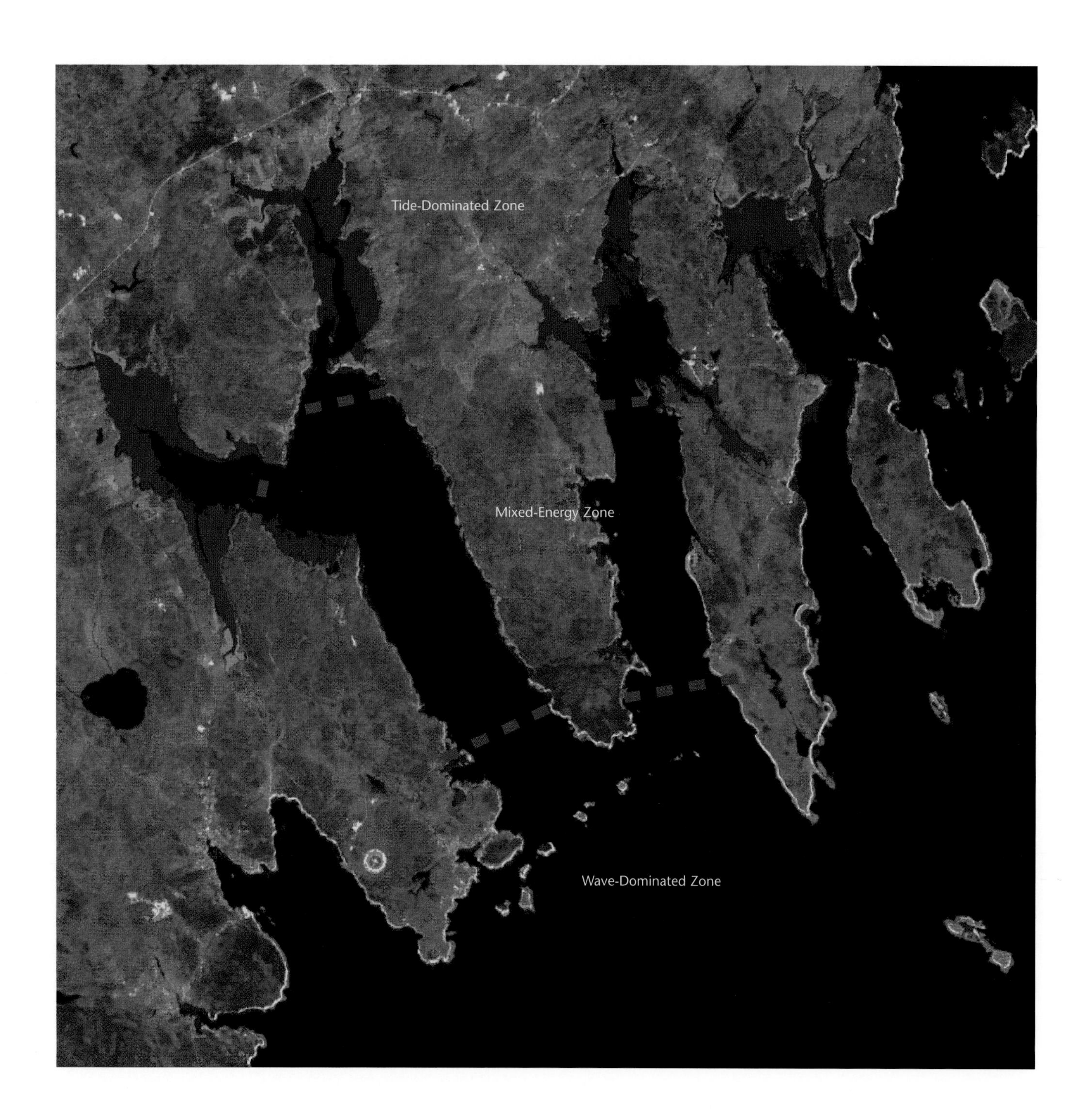
Tide-Dominated Zone
Mixed-Energy Zone
Wave-Dominated Zone

3.13 Embayment Structure
(left)

In the tide-dominated zone, salt marshes and mudflats (dark blue) flourish. In these quiet, more protected areas rivers deposit much of their nutrient-rich sediment load. In the mixed-energy zone the process of erosion carries away glacial deposits. In the high-energy, wave-dominated zone all glacial deposits are gone. Below the water on the seaward side can be found evidence of former salt marshes, indicating the steady advance of the sea landward. *Credit: R. Podolsky, S. Meyer, Island Institute; GAIA image*

3.14 Sea Level Rise Mosaic
(right)

The images presented here depict four stages of coastal inundation, simulating the process that occurred throughout the Gulf of Maine following the melting of the last ice sheet. This mosaic of images was created by digitally "flooding" an elevation model of the region, with each image showing the effect of approximately a 75-foot (23-meter) rise in sea level. Although the simulations assume that coastal elevations remain constant, we know that land rebounds concurrently with sea level rise following the melting of a glacier. Therefore, the actual changes in shorelines were more complex and dynamic than can be shown here. Nevertheless, the simulation dramatically demonstrates the scale of changes in coastal geomorphology in the Gulf of Maine that occurred following glacial retreat, and the effects of rising seas on the arrangement of islands and configuration of shorelines. *Credit: R. Podolsky, Island Institute*

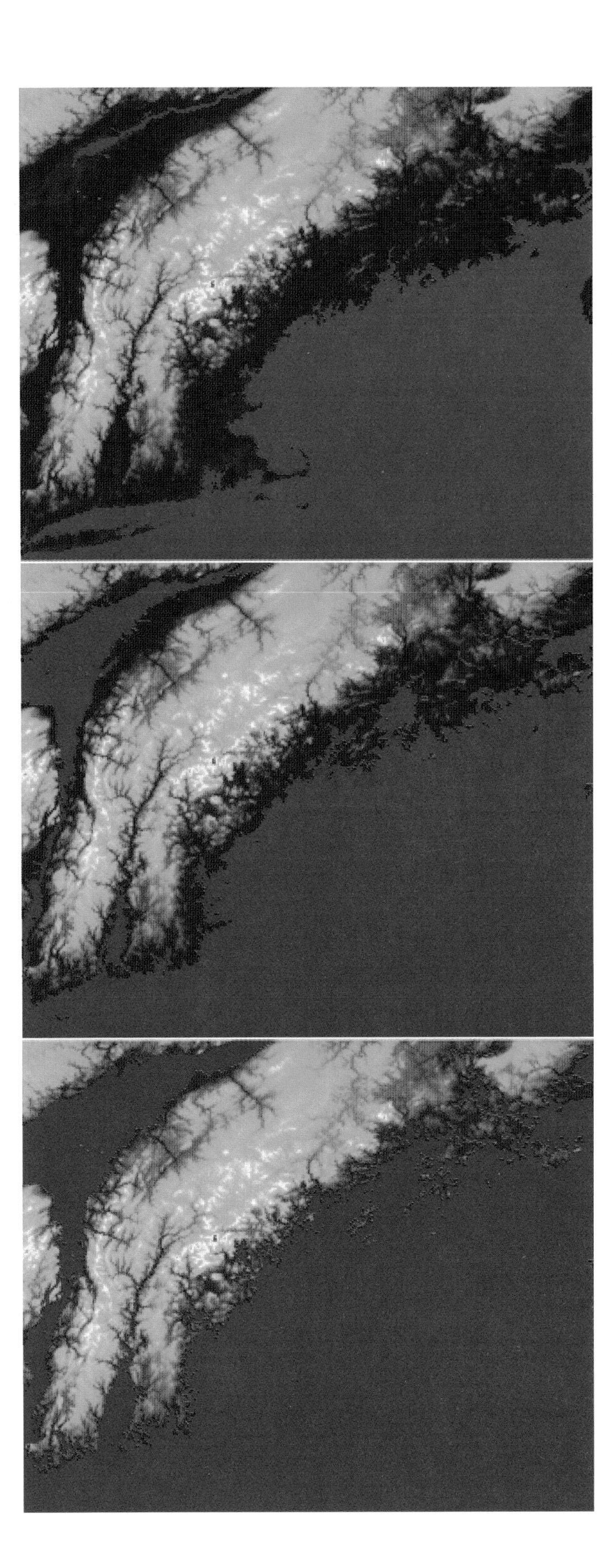

shape of the coast has witnessed shoreline change. Peninsulas have become islands and islands have become submerged shoals. Most of these changes have come with little public concern, because, historically, people have simply moved to accommodate the rising sea. It has only been in the past century, and especially during the last few decades, with people moving in such large numbers onto the coast, that retreat has become difficult. Simultaneously, it has been during this century that we have recorded the most rapid rise in sea level in several thousand years, a rise that may be partly attributed to our own burning of fossil fuels (figs. 3.14, 3.15).

There is little we can do to prevent the rise of the sea. Many properties built too close to the ocean will be annually claimed by ocean storms trying to push our beaches landward. Where seawalls have been erected to hold back the sea, millions of dollars will be required to perpetually maintain the engineering structures. This may be possible at places like Boston, where most of the city rests on the filled remnants of tidal flats and salt marshes. Such highly populated low areas will be walled and drained by pumps just as New Orleans, Venice, and the Netherlands are today. Less extensive development, such as exists along our beaches and eroding bluffs, however, may not be protected (fig. 3.16). The costs will be too much for individuals to endure, and the loss of productive coastline too much for the public. A day of reckoning is coming for each location along the coast of the Gulf of Maine when a decision must be made to armor the shore or to retreat landward. The costs of engineering, and the associated aesthetic and ecological losses of halting the main process that formed the present Gulf of Maine, will undoubtedly lead to a wiser strategy to accommodate the sea than has existed for the past few decades.

3.15 Pleistocene Islands: Penobscot Bay 10,000 Years Ago

The most significant geological process that has affected the Maine coast and islands in recent geologic time is the rise and fall of the ocean brought about by glaciation. Because so much of the earth's water was tied up in ice during glaciation, sea level fell by as much as 300 feet (almost 100 meters). During periods of glacial melt (interglacial periods), the process went in reverse: the earth warmed, glaciers melted, and sea level rose. This glacially induced rise and fall of the sea is a global phenomenon that can dramatically rearrange the world's coastal regions.

How has the rise and fall of sea level in the Gulf of Maine affected the Penobscot Bay coast and islands? To visualize the impact of this very recent rise in ocean level one must imagine what the Maine coast would look like with several hundred feet of ocean removed. We examined bathymetric charts and ocean floor contour lines that identify past shorelines and former islands at 180 feet (55 meters) below present sea level. The map shown here was produced by this method and traces the impact of sea level rising over the last 10,000 years of geological time. The map presents four phases, each as a conceptual snapshot of how the Maine coast and islands probably appeared at some period during the last 10,000 years. Precise dates cannot be attached to these past coastal configurations, because as the sea has risen there has been a concurrent glacial rebound with the land rising in response to the removal of the weight of the glacier. The weight of the immense ice sheet that covered the Gulf of Maine out to Georges Bank was enough to warp the earth's crust down along the entire length of the Maine coast.

This map reveals some surprising facts. Approximately 10,000 years ago, when the ocean was 180 feet lower, the mainland extended much farther out to sea and absorbed many Penobscot Bay islands. Inshore islands in shallow water of less than 180 feet probably became part of the mainland. Offshore islands in water deeper than 180 feet became larger islands. And submerged seamounts, banks, and ledges with elevations exceeding 180 feet became new islands. Large rivers meandered through an additional 10,000 square miles of mainland, gliding past coastal hills with names like Monhegan and Hurricane.

What groups of plants and animals would have been most affected? What happened to early human settlers on the coast? According to archaeologist Arthur Speiss, the majority of precolonial archaeological sites on the Maine coast and islands may already be subtidal. This indicates that there was a significant amount of human activity at a time when the coast and islands had a very different configuration. It also indicates that we must think carefully about

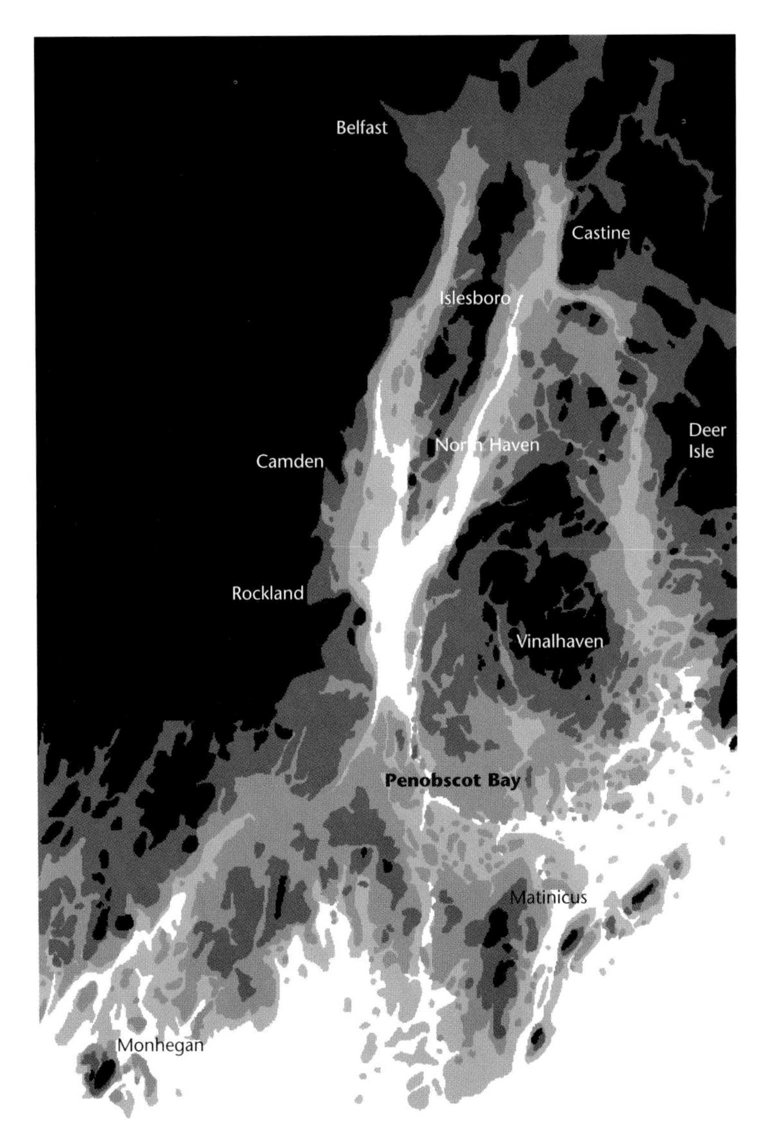

preserving for study the remaining above-water sites before erosion destroys them, as well as formulating a plan to study the already submerged sites.

Oceanic birds such as storm petrels, terns, puffins, auks, and gulls require mammal-free nesting islands to prevent adults and young from being eaten. The lowering of the sea level connects formerly predator-free nesting islands to the mainland by way of land bridges. Land bridges would allow the arrival of predatory mammals, and colonies of seabirds would be quickly destroyed. Leach's storm petrel is a pelagic seabird currently known to nest on only 17 Maine islands. Its nesting islands are all predator-free and are among the few Maine islands that have not been connected by land bridges to the mainland in over 10,000 years.

The red-backed salamander is another animal whose present distribution may be the result of a past coastal configuration. Like most amphibians, it does not disperse easily over salt water, so its abundance on many Maine islands is problematical. One possible explanation is that these salamanders represent relict populations isolated on islands when rising seas flooded mainland peninsulas. Other relict populations may include the gray squirrel and heavy-seeded tree species such as oaks and beeches.

Even today, the interaction of the land and sea along the Maine coast is best viewed as a dynamic relationship. Along different parts of the coast, land is subsiding and the sea is consequently advancing inland. This change of the height of the land with respect to the sea is measured in inches per century in midcoast Maine but in feet per century in Cobscook Bay where crustal activity is greater. Although changes in land and sea measured in inches per century may not seem like much, when they occur along unstable bluffs or are combined with storm surges the results can be catastrophic.

Managing our coast and islands with an eye toward these long-term geological events does not preclude wise short-term planning. A geological perspective brought to the planning process will help assure that future generations do not inherit environmental problems that we could easily avoid today. *Credit: graphic: S. Meyer; data: R. Podolsky, W. Marshall*

3.16 Sea Level Rise, Student Project

This student project shows simulated sea level rise of 1, 2, and 10 meters from present-day conditions in Saco Bay, Maine. Although no one expects the sea to rise to these levels in a human time scale, sea level has risen much more dramatically than this in geologic time scales. In the words of the student, "I was looking for a process that would show sea level rise on an image. I used a lot of maps to interpret land areas that would get covered by water as sea level rises, like wetlands and marshes. And I went ground truthing. I used USGS maps to figure out what land contours would also get covered as the sea level rose. I learned that this is what I want to do for a career." *Credit: J. Everette, grade 9, South Portland High School; teacher, J. Salisbury; courtesy of Gaia Crossroads Project, Bigelow Laboratory for Ocean Sciences*

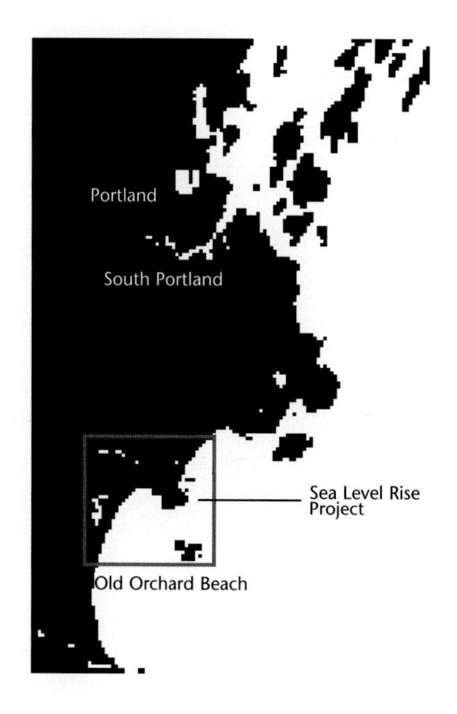

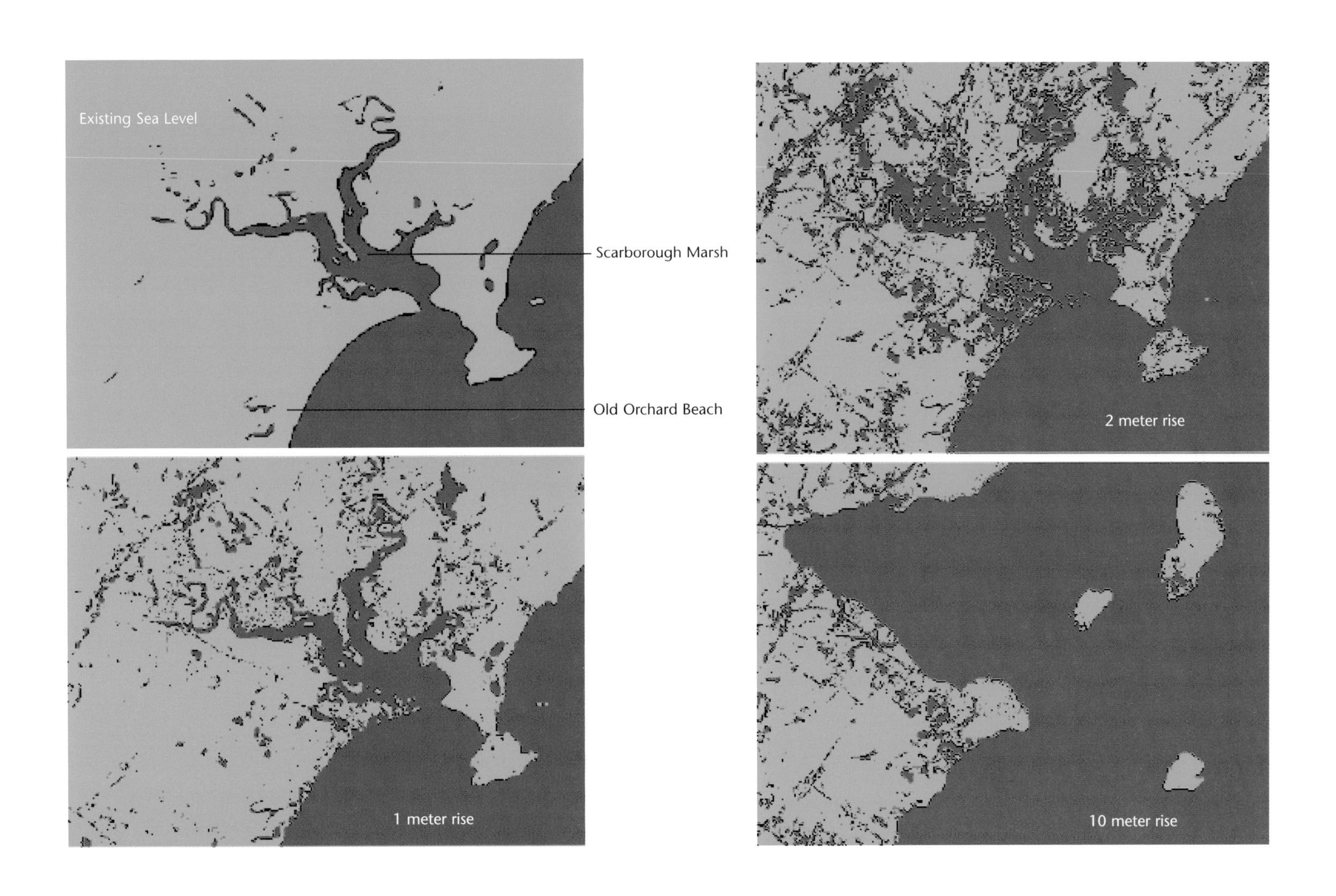
Existing Sea Level
Scarborough Marsh
Old Orchard Beach
2 meter rise
1 meter rise
10 meter rise

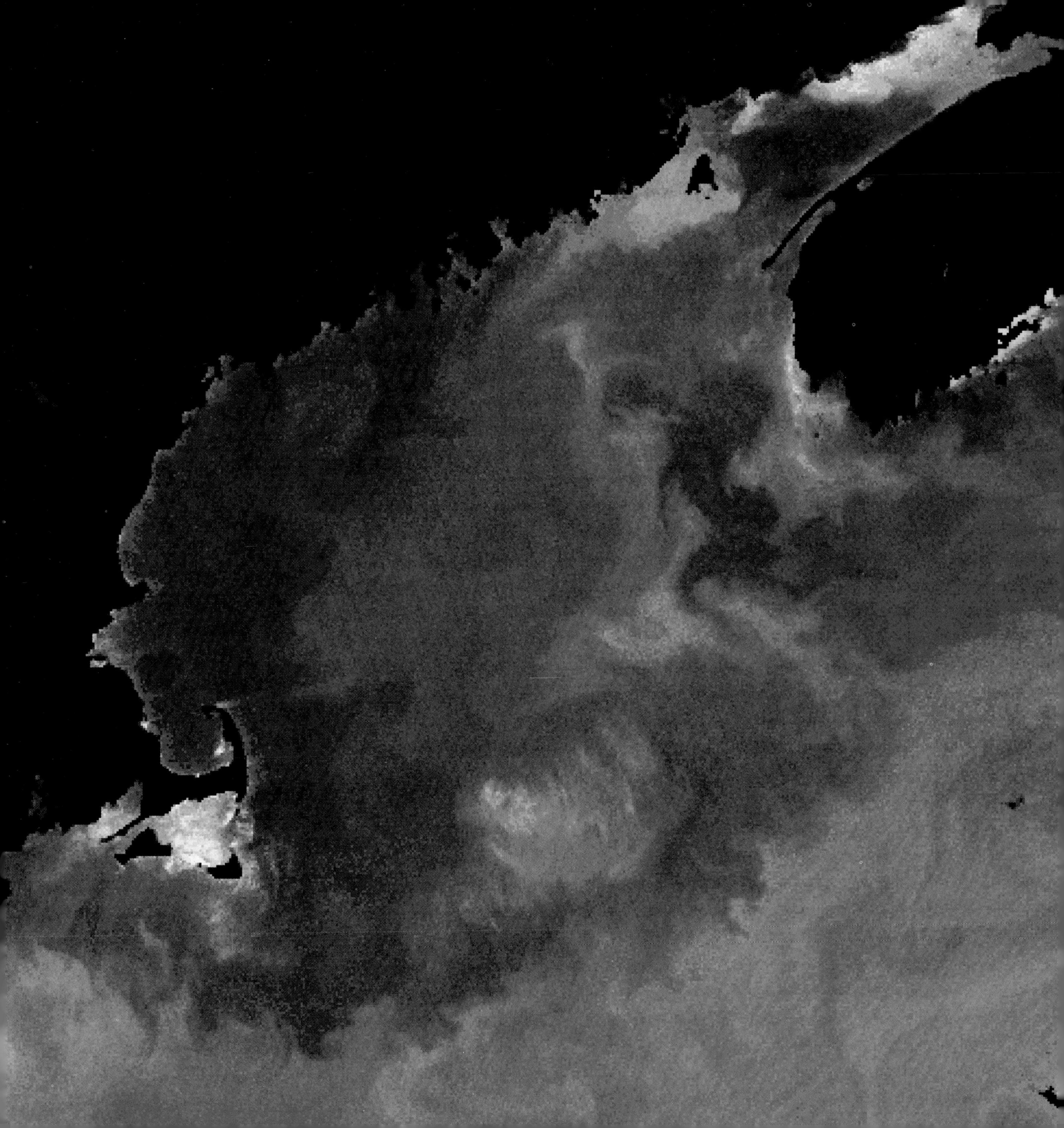

4

THE GARDEN IN THE SEA

BIOLOGICAL OCEANOGRAPHY

Charles S. Yentsch, Janet W. Campbell, and Spencer Apollonio

4.1 Ocean Colors of the Gulf of Maine

The Coastal Zone Color Scanner (CZCS), launched aboard the Nimbus 7 satellite in 1978, was one of the earliest satellite sensors designed to collect oceanographic data. This image from June 14, 1979, shows variations in reflected sunlight in the garden of the Gulf of Maine. The areas of greatest turbidity (in yellow) are found in the upper Bay of Fundy and in Nantucket Sound. Phytoplankton blooms (gray-green areas) appear dark because the phytoplankton absorb sunlight and hence less light is reflected from the water. High concentrations of phytoplankton in this image are found near the mouth of the Bay of Fundy off Grand Manan Island, along the eastern coast of Maine, and in the island-studded bays of Maine, including Machias, Frenchman, East and West Penobscot, Muscongus and Casco bays. In Massachusetts, the waters off Cape Ann and in Massachusetts Bay are similarly enriched. The central portion of Georges Bank, where tidal mixing is most intense, appears turbid, whereas the outer fringes are high in phytoplankton. Less productive waters of the Atlantic (blue) are clearer and reflect more light. *Credit: image: J. Campbell, Bigelow Laboratory for Ocean Sciences; data: G. Feldman, NASA Goddard Space Flight Center*

Colors of the Gulf

The Atlantic Ocean and the Gulf of Maine are separated by giant sand dunes that rise from the ocean floor to within a few tens of meters of the surface. From shore, the ocean lies far beyond the horizon, perhaps 200 or more miles (more than 300 kilometers) offshore. These submerged topographic features—Georges Bank and Browns Bank—are not visible from the headlands nor from ships, yet their effect on surface waters is clearly visible from space.

In a Coastal Zone Color Scanner (CZCS) image (fig. 4.1), one can almost imagine that the sand dunes are visible over the top of Georges Bank, and indeed the ridgelike patterns in the image are aligned with the ridges and troughs of the dunes. Most of the light recorded by the CZCS instrument is reflected from the upper 10 meters of the water column. Since the peaks of the dunes are generally deeper than 10 meters, it is unlikely that the dunes themselves are visible. But oceanographers believe that the dunes set up a pattern of cyclical vortices that stir bottom material into the surface waters. Rich in organic debris, this bottom material is the fertilizer that makes Georges Bank one of the richest fishing grounds in the world.

Headed out into the Atlantic from the Gulf, a shipboard observer might notice a color change from dark blue to clearer blue water. Such differences can be enhanced in satellite images of ocean color. In a color-composite image (e.g., fig. 4.1), the Atlantic Ocean appears bright blue because it is dominated by blue light reflected from clear oceanic waters. Inside the Gulf of Maine there is a small area of bright blue as well—perhaps Atlantic water drawn into the Gulf through the Northeast Channel. Most of the Gulf, however, is a darker green or gray-green color, indicating that less blue light is being reflected. (When observed from high altitude, the water of the Gulf of Maine appears dark green, contrasting with the azure blue of the North Atlantic.) A student of light and color would realize that something is absorbing most of the blue light entering the Gulf water, making it look different from the open ocean. In these areas the blue light is being absorbed by tiny floating cells of marine algae, called *phytoplankton*, and thus less is reflected back into space. Understanding phytoplankton is the key to understanding the unique properties of the Gulf of Maine and Georges Bank (fig. 4.2).

From detailed biological and optical studies of waters in the Gulf of Maine we know that these tiny plants, like the grasses on land, absorb sunlight and turn the sun's energy into organic material through photosynthesis. In the process they create a rich food supply for plant-eating animals, and as a by-product they produce oxygen. The scientific term for this fundamental and critical process is *primary production.*

The biology of primary production interacts with the physical and chemical features—tides, temperature, salinity, and other factors in the Gulf of Maine and on Georges and Browns banks—to create a diverse and abundant supply of marine life. The Gulf of Maine is, in effect, an aquatic garden, based on tiny

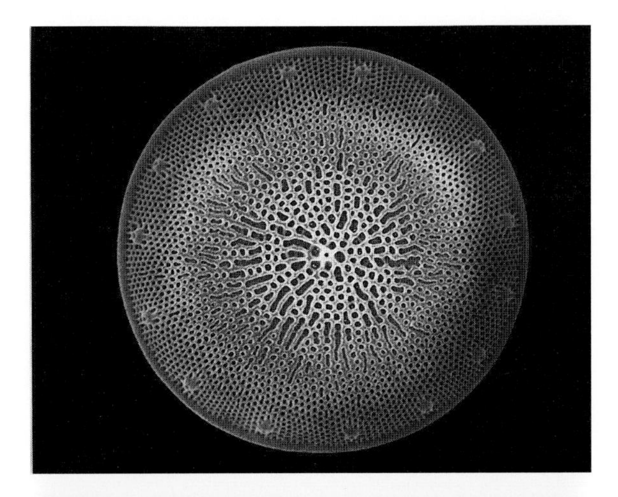

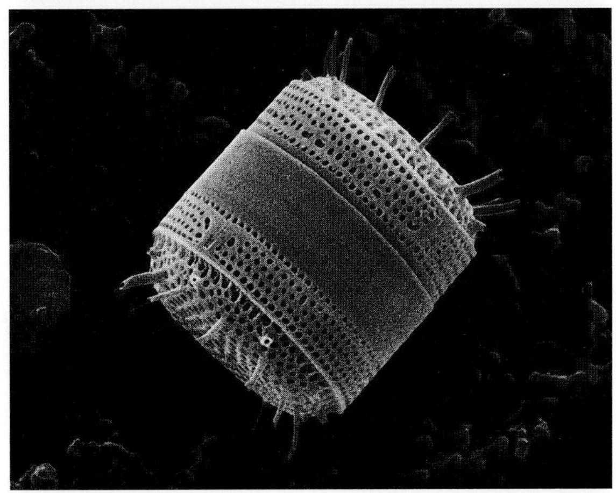

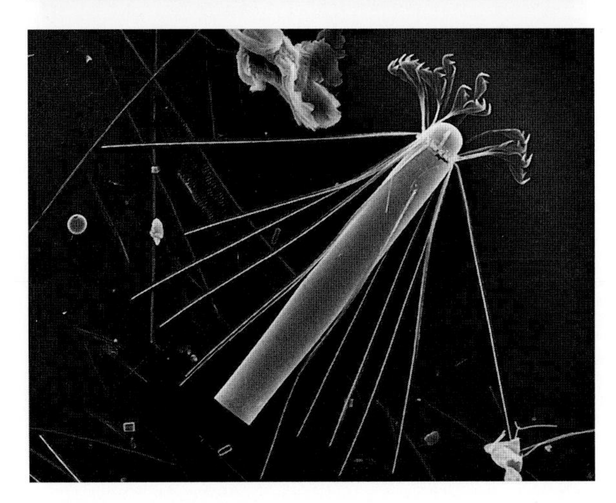

4.2 Phytoplankton Close-up

Various views of phytoplankton, the single-celled plants that are the foundation of the marine food web, equivalent to grasses and trees on land. These particular plankton specimens are all diatoms. *Credit: D. Breger, Lamont Geological Observatory*

green plants so small that you cannot see them with the naked eye, on which a stunning variety of marine species depend, which in turn have fed humans for generations.

We tend to think of the oceans of the world as representing a vast biological potential. But this is a myth. The oceans of the world are deserts, by and large, because their nutrients are too scarce and scattered to support much planktonic life, the basis of all marine food webs. But in the Gulf of Maine, especially on its seaward-facing banks, nutrients are constantly stirred into the upper, well-lit zones of the water column, fertilizing the immense green growth of tiny plants. The fundamental productivity of these waters is what fed the region's earliest human inhabitants, attracted early fishermen and explorers, built settlements, and sustained generations of coastal residents scattered around its rim.

The productivity of the Gulf of Maine—meaning its ability to capture sunlight and transform it into floating plant life that all other species depend upon—though not as great as in the adjacent Georges Bank, is still high compared to most other ocean regions of the world (fig. 4.3). There is a rich store of nutrients in the deep waters of the Gulf. They are continuously replenished by nutrients in the waters that flow inward from the North Atlantic and westward from the continental shelf east of Nova Scotia, through the Northeast or Fundian Channel. While in the Gulf of Maine they are mixed by the wind-driven turbulence and seasonal convection of each winter.

The Gulf is washed by cold waters, which hold a greater amount of dissolved oxygen and carbon dioxide than can warmer water. This is an important factor in the Gulf's enhanced productivity. The drainage of many rivers contributes an additional abundance of nutrients. On average, 250 billion gallons (950 billion liters) of fresh water empty into the Gulf each year from more than 60 rivers. The numerous estuaries located around the mouths of these rivers also function as important breeding and feeding grounds for many fish and shellfish populations of the Gulf. The natural productivity of the Gulf of Maine itself is supplemented by the rich productivity of Georges Bank, some of which is exported into nearby parts of the Gulf. Many species migrate into the Gulf of Maine in order to feed upon that abundance of food.

The Heat of the Sun, the Pull of the Moon

Within the Gulf of Maine, sea surface temperature, related to water depth, is a critical factor influencing the rate of growth of phytoplankton, upon which so much marine life depends. In summer the coolest waters are located over Georges Bank and other shoal areas, where winds and tides mix the sun-warmed surface waters with cooler waters from below. Over deep basins of the Gulf—Wilkinson, Jordan, and Georges basins—the water is more stable, and thus a layer of warm, buoyant water floats near the surface during the warmer months.

Warmed by the sun, an unperturbed Gulf of Maine would become layered (stratified) in terms of the distribution of heat. The surface waters, heated by sunlight, would become less dense than unlighted cold, deep water and thus tend to remain on top. Water masses where warm surface layers overlie cold layers are termed stable; that is, they don't mix unless other forces are brought to bear. Wind blowing across the surface can disrupt the stability and mix warm and cold water, as can currents induced by tides.

These stabilization and destabilization cycles have a marked effect on the growth and abundance of phytoplankton. The growth of phytoplankton requires light and specific nutrients such as nitrogen and phosphorus. In a stable water mass, the sunlit surface layers become stripped of these nutrients because the nutrients gradually sink to the sea bottom and are unavailable for use. Consequently, phytoplankton growth is markedly reduced where stable water columns occur. But with destabilization or vertical mixing, these nutrients are recycled from below and growth improves rapidly, resulting in a bloom of phytoplankton.

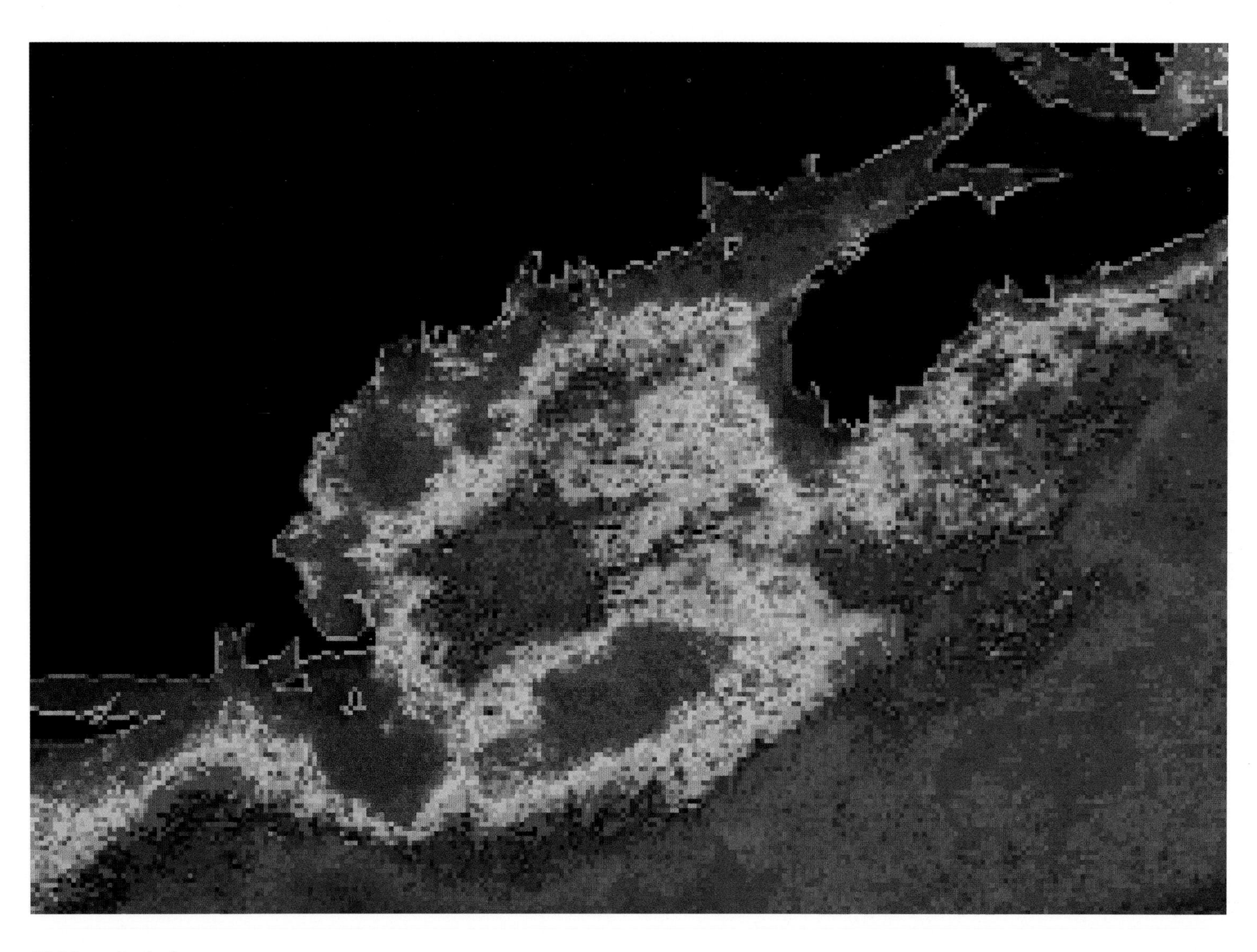

4.3 Primary Production

This Coastal Zone Color Scanner image of chlorophyll distribution shows the zones of highest productivity (dark red) in the Bay of Fundy, along the Maine coast, and over Georges Bank, underscoring the unique enclosed nature of this marine region. *Credit: G. Feldman, NASA Goddard Space Flight Center*

Where do the nutrients that make up this aquatic garden come from? From the waters of the Atlantic, for one thing, via the two channels—Northeast Channel and Great South Channel—that connect the interior of the Gulf to the Atlantic (fig. 4.4). If not for these two channels, Georges and Browns banks would almost isolate the Gulf of Maine from the rest of the Atlantic Ocean. Nutrient-rich water comes through the deep Northeast Channel, and surface water flows out via the shallower channels to the south.

The interior of the Gulf is made up of two basic water types, each with three distinct layers—surface, middle, and bottom layers. The combination of inflow and outflow of water through Northeast and Great South channels, along with variations in surface heating and freshwater inputs from the vast river drainages, maintains the continuity of the marine system in the Gulf. But there is not enough energy in this inflow-outflow pattern alone to provide adequate vertical mixing to sustain complex food chains. In the Gulf of Maine, it is the effect of our strong tidal currents that makes up the difference.

Again, it is important to stress the role of topographic features, mainly Georges Bank and associated shoal regions, in shaping the Gulf of Maine's water temperature and productivity regimes. The shallowness of water over these shoals increases the amplitude of the ocean tidal wave as it is drawn into the Gulf of Maine by the pull of the moon. On the seaward side of Georges Bank the height of the ocean tidal wave is approximately three feet (a meter). As it moves over the bank, its height significantly increases and strong tidal currents develop. As these currents move across the shoals, the friction of the bottom causes the water column to mix. Tidal action sustains this condition. We can view the productivity of the Gulf of Maine as an engine requiring two sources of energy: sunlight for photosynthesis, and the pull of the moon (i.e., tidal energy) to keep the water masses stirred.

Beginning in 1912, Henry Bigelow, founder of Woods Hole Oceanographic Institution, made extensive measurements of water temperature and salinity from the sailing schooner

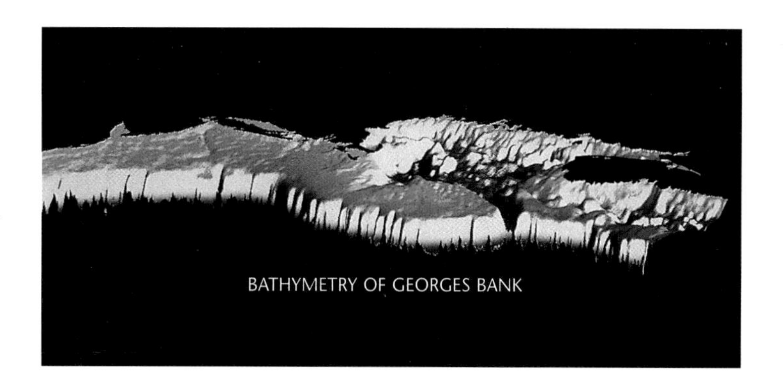

4.4 Bathymetric Terrain Model

This illustration compiled from underwater radar data shows Georges Bank as it might appear from offshore in the Atlantic Ocean. Note the deep Northeast Channel separating Georges on the left of the image from Browns Bank on the right. *Credit: courtesy of Bigelow Laboratory for Ocean Sciences; P. Larsen*

4.5 Henry Bigelow
(left)

Henry Bigelow, the father of modern oceanography and the first director of the Woods Hole Oceanographic Institution, spent most of his life collecting data on the currents, plankton, and fish of the Gulf of Maine. In his early years, he collected data from the decks of his schooner *Grampus. Credit: photo courtesy of Woods Hole Oceanographic Institution*

4.6 Sea Surface Temperatures
(right)

Satellite data can be used to detect relatively slight differences in sea surface temperatures. In this image shades of blue show the distribution of the coldest waters in the Gulf and nicely outline the area over Georges Bank southeast of Cape Cod. Cold water is also located over Browns Bank off the tip of Nova Scotia and appears to trend from the tip of Grand Manan Island southwest along the coast of Maine. Light green areas show the warm waters of the Gulf Stream along the bottom of the image, tendrils of which appear to be filtering in on the southern edge of Georges Bank. Warm waters also appear widely distributed over the western half of the Gulf of Maine. *Credit: data courtesy of C. Yentsch, Bigelow Laboratory for Ocean Sciences; image enhancement: T. Ongaro, P. Conkling, Island Institute*

Grampus (fig. 4.5). From these measurements he was able to map areas of high and low stability. He concluded that areas of high productivity and low stability occur in regions where tidal currents are strong. In the central basin of the Gulf, water masses are stable, while the well-mixed, less stable waters are over Georges Bank and other shoals. Because of this mixing, the surface temperature of the mixed waters will be lower than those of the stable water masses.

Today, satellite imagery that senses temperature differences on the ocean surface can provide thermal signatures for the entire region. There is a strong resemblance between the pattern of temperature and the topographic outlines of the banks. Patterns of cold and warm surface temperature (fig. 4.6) are persistent features in the Gulf of Maine during summer months. Satellites only record the temperature of a thin surface layer, but

from shipboard measurement we can see that the surface measurements closely represent the temperature for the entire water column over Georges Bank, illustrating that these waters are mixed from top to bottom. In contrast, the surface waters of Wilkinson Basin, northeast of Cape Cod in the Gulf of Maine, are considerably warmer than deeper layers, demonstrating that they are not being mixed.

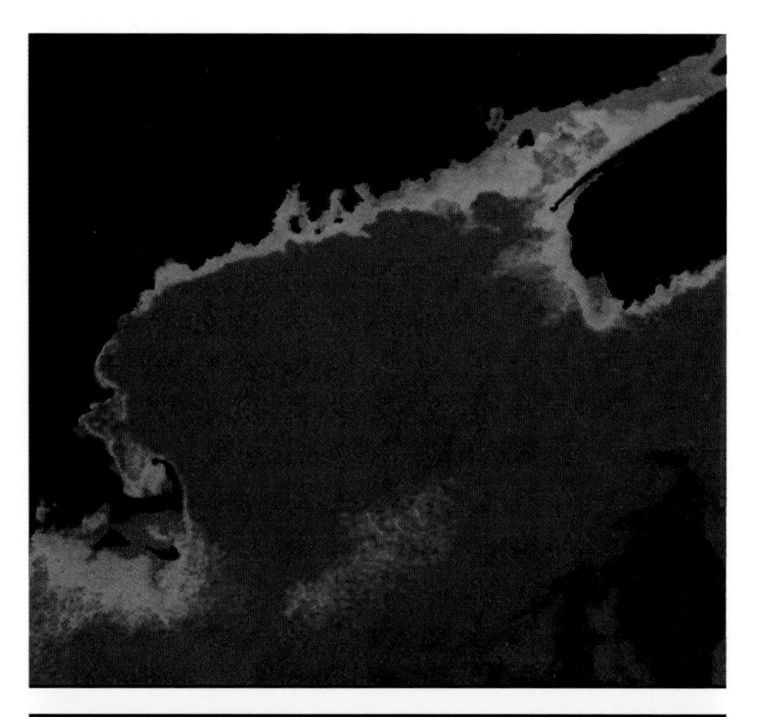

Four Seasons in the Garden

The sea in general is not a gardener's delight, especially during the summer months. Things seem to be in the wrong places. The fertilizers necessary for plant (phytoplankton) growth are often located in deep, cold waters where the light required for photosynthesis is weak. Sunlight is strongest in surface waters where nutrients are sparse. As described earlier, we expect well-mixed cold-water regions to be rich in nutrients, and warm, stable water masses to be poor in nutrients.

We can turn to satellite imagery—a satellite sensor that measures the color of the ocean, reflecting the abundance of phytoplankton and primary productivity—to see if this is true. The first step is to compare the patterns of color with those for temperature. This type of comparison clearly shows that the cold, mixed waters are five times richer than the stable water masses occupying much of the central basins of the Gulf. The well-mixed areas account for 60 percent of the total phytoplankton production while occupying only about 30 percent of the total area of the Gulf.

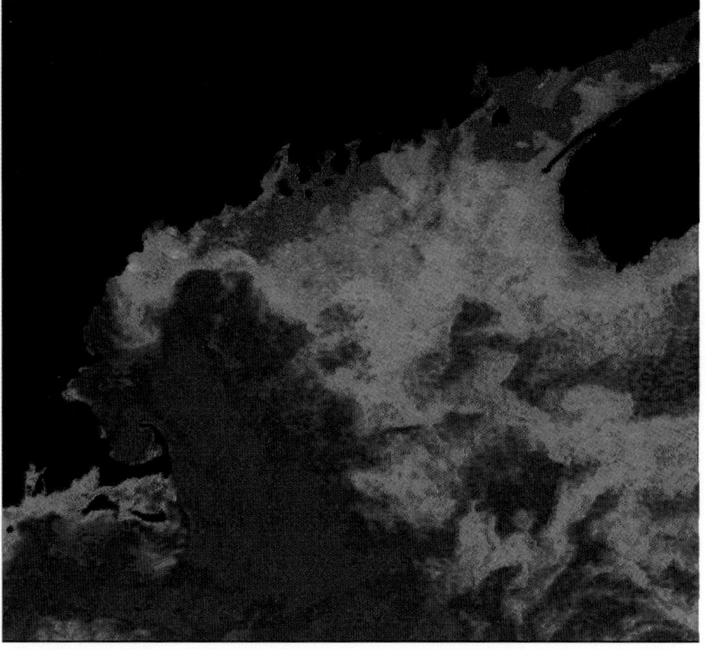

The energy for such mixing as takes place in the water masses of the central basins comes from winds, and from the heating and cooling of the sea surface. The surface waters in these basins reach their maximum temperature toward the end of August. Then, as the amount of sunlight declines, these layers begin to cool, especially during periods when cold continental air masses overlie the Gulf. The cooling causes the surface waters to become more dense and to sink, a process referred to as *convective overturn.* Northerly winds accelerate this process.

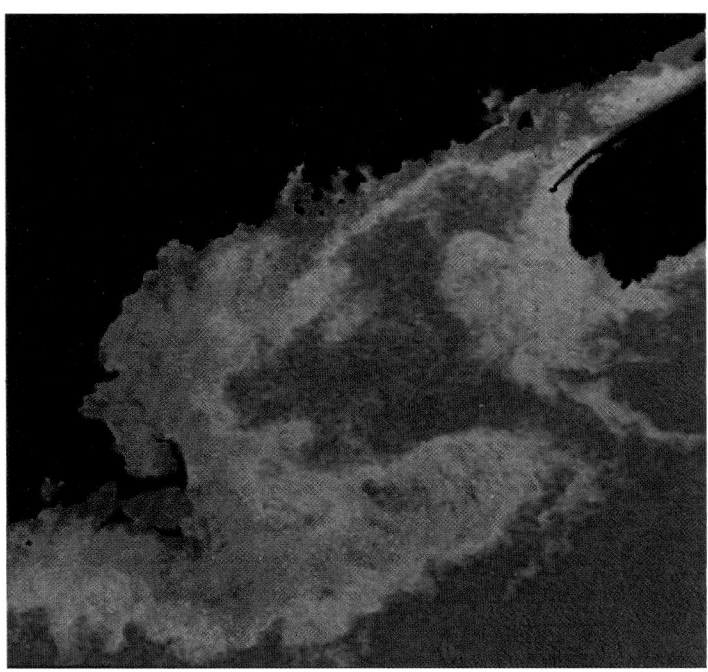

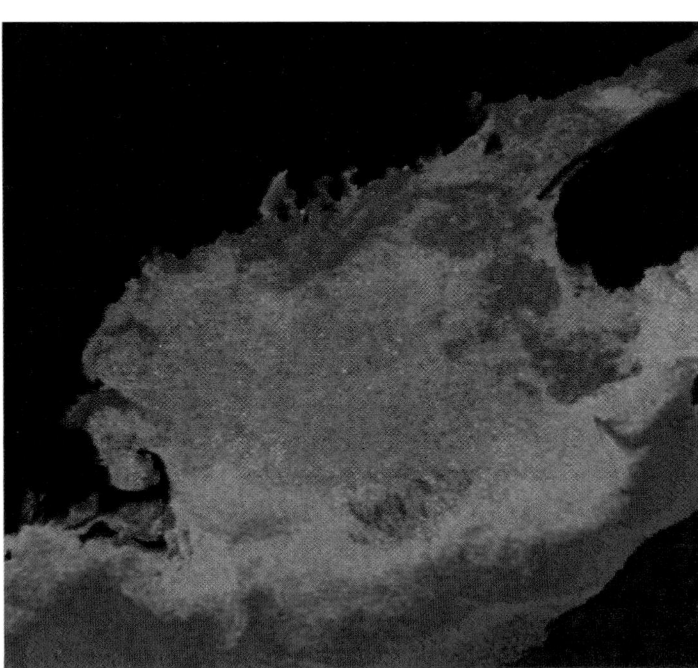

4.7 Phytoplankton Bloom

This series of Coastal Zone Color Scanner (CZCS) images of the Gulf of Maine shows the sequence of the 1979 phytoplankton bloom from March 22 to May 8, June 14, and August 31. In the spring of each year, temperature stratification, sunlight, and nutrients allow for a vast bloom of chlorophyll in a warm layer near the top of the water column. The presence of chlorophyll is detectable by CZCS satellite, which picks up light reflected from the surface layer. By summer, the phytoplankton community has used up all the nutrients that have sustained it and essentially starves. In the fall another chlorophyll bloom will occur that is dependent on nutrients from upwellings. *Credit: C. Yentsch, Bigelow Laboratory for Ocean Sciences*

The immediate benefit of the sinking of the surface water is similar to what happens when we turn over a garden: the nutrients from the deeper layers are brought to the surface. But when winter approaches, the amount of sunlight decreases markedly. As these layers are overturned, the phytoplankton is mixed deep into the water column and hence receives even less light. Consequently the rate of growth is retarded.

The depth at which phytoplankton grows is shallowest during winter, when the sun is low. When the sun is high in summer, the growth area (which oceanographers call the *eutrophic zone*) is deep. The combination of a deep convective mixed layer and shallow eutrophic zone is not good for growth because the deep mixing separates phytoplankton from the zone of sunlight. But when areas of sunlight and vertical mixing overlap, such as during April-May and September-October, large blooms of phytoplankton are observed in the central basins of the Gulf (fig. 4.7).

The important factor in this seasonal scheme is the extent to which the depth of convective mixing exceeds the depth of the eutrophic zone. Again, the importance of the banks is that their shallow depth means that vertical mixing occurs throughout the water column.

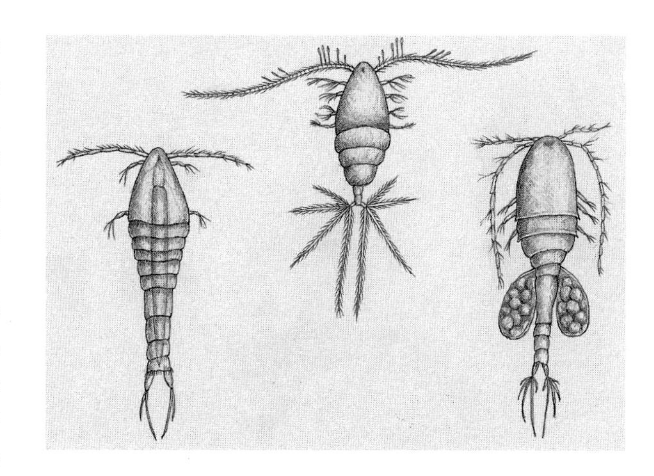

4.8 Zooplankton

Three species of copepods (from left, *Calanus, Canthocamptus,* and *Cyclops*), which are tiny floating crustaceans that make up a great portion of the zooplankton in the Gulf of Maine. *Credit: K. Bray*

The Role of Zooplankton

All fish and shellfish of the Gulf of Maine and Georges Bank are ultimately dependent on the basic productivity of marine phytoplankton at the bottom of the food chain. But there is an intermediary category of marine life—*zooplankton*—that serves to transfer the sun's energy captured by phytoplankton to other species of fish and shellfish (figs. 4.8, 4.9).

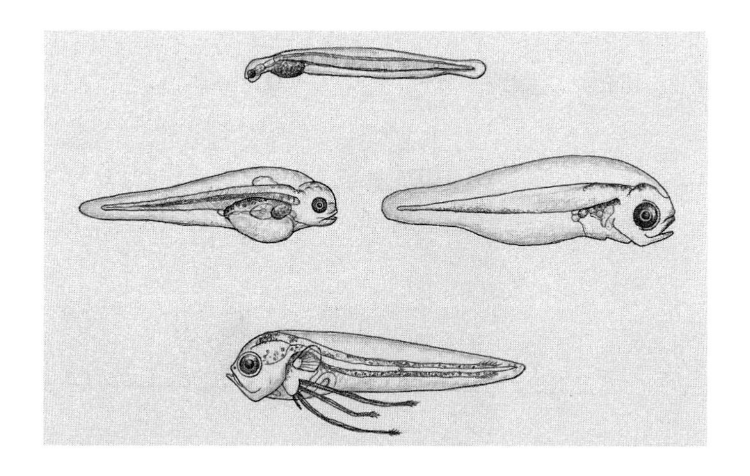

4.9 Larval Fish

Fish larvae, which swim through the water column feeding on plankton, are also a part of the zooplankton in the Gulf of Maine. Larval herring, haddock (middle two), and cusk are shown here. *Credit: K. Bray*

Zooplankton comprises those tiny crustaceans, worms, mollusks, and fish larvae that move up and down in the water column feeding on the microscopic marine plants. Their patterns of abundance and distribution vary in predictable ways along the Maine coast because of coastal oceanographic patterns. The abundance of zooplankton in the Bay of Fundy is markedly less than in the rest of the Gulf of Maine because of the great tidal action and the great turbidity, a result of the stirring and mixing of bottom sediments caused by that tidal action. Turbidity greatly reduces the growth of light-dependent phytoplankton. The Bay of Fundy simply doesn't have the basic production capacity to support an abundant zooplankton population.

But elsewhere in the Gulf of Maine, especially in its southwestern regions, zooplankton populations develop and feed on the soupy green broth of phytoplankton that has bloomed earlier.

The Mystery of Coccolithophore Blooms

Coccolithophores are a group of phytoplanktonic organisms characterized by a covering of platelets known as coccoliths (fig. 4.10)—round, ornate structures made of calcium carbonate (chalk). Coccolithophores are always present in the Gulf of Maine, but sometimes an extensive bloom occurs such that the coccolithophores far outnumber all other species. During these episodes—lasting up to a month or so—coccolithophore cells grow and divide as frequently as twice a day. In addition, they produce and continually shed their coccoliths. The waters become filled with live coccolithophore cells, and for each cell dozens of loose coccoliths are also floating in the water. Because the coccoliths are made of a chalklike substance, they are white and can be seen as bright areas in satellite imagery.

"White waters" have been reported by mariners for centuries. The first satellite images showing a bloom in the Gulf of Maine were obtained in 1983. At that time, however, the bloom was unnoticed, and no ships were dispatched to study it. In fact it wasn't until several years later that an archived satellite image was analyzed and the bloom discovered. Beginning in 1988, a team of scientists at the Bigelow Laboratory for Ocean Sciences in West Boothbay Harbor prepared to study a bloom they thought might develop in the Gulf of Maine in early June.

The scientists acquired daily satellite images and inspected them to determine when the bloom was occurring (fig. 4.11). At the first sight of the bloom, a research vessel was dispatched to the area of maximum intensity and then guided along routes transecting the bloom in several directions. At its maximum, the bloom covered nearly 15,000 square miles (40,000 square kilometers). Another large bloom occurred in 1989, but since 1989, there have not been any extensive coccolithophore blooms in the Gulf of Maine. A much smaller, more localized bloom occurred in 1990, and there was none in 1991. Why they occur in some years and not in others remains a mystery. What triggers them? Are blooms becoming more common? We don't know the answers because satellite technology has only recently enabled us to know when and where blooms occur. Before we had the ability to observe them with remote sensors, many blooms may simply have gone unnoticed.

Coccolithophore blooms are a natural part of the productive processes in the Gulf of Maine. Their uniqueness lies in the detectability of their chalky plates, and thus their ability to show us the timing, growth, and dispersion of a bloom of microscopic plants and relate this understanding to the Gulf as a whole.

Our Young Gulf

The age of the Gulf of Maine is frequently put at about 16,000 years, approximately the time since the most recent glaciation retreated from our area. This is very young when compared to the age of most geographic features on the surface of the earth. The Atlantic Ocean, for example, is about 150 million years old. But the *functional* age of the Gulf is even less than 16,000 years; in fact, it is probably about 4,000 years, less than the duration of recorded human history and more recent than the arrival of early man in the area we now call the Gulf of Maine watershed.

4.10 Coccolithophores Close-up
(left)

Researchers at Bigelow Laboratory are investigating the role that coccolithophores play in the oceanic carbon, carbon dioxide, and sulfur budgets, as well as the effects that coccolithophores have upon the formation of the surface mixed layer of seawater. *Credit: D. Breger, Lamont Geological Observatory*

4.11 Coccolithophore Blooms
(right)

Using data collected by satellites, scientists at Bigelow Laboratory are studying the occurrence and distribution of phytoplankton in the Gulf of Maine. These images were recorded June 12, 1989, by a NOAA satellite sensor. The images have been enhanced to reveal water features associated with a coccolithophore blooms (white areas). Land and clouds appear black.

Coccolithophore blooms form within stratified surface waters during the late spring and early summer over large areas of the Gulf of Maine and North Atlantic. Unlike other groups of phytoplankton, these organisms have unique optical properties due to their external plates composed almost entirely of calcium carbonate. The plates greatly increase light scatter, which increases the rate of absorption, which in turn enhances the rate of heating within surface waters. Stratification of the water column is a prerequisite for the formation of coccolithophore blooms; once developed, the coccolithophores appear to further stratify the surface waters where they form. *Credit: image: S. Ackleson, Bigelow Laboratory for Ocean Sciences; data: G. Feldman, NASA Goddard Space Flight Center*

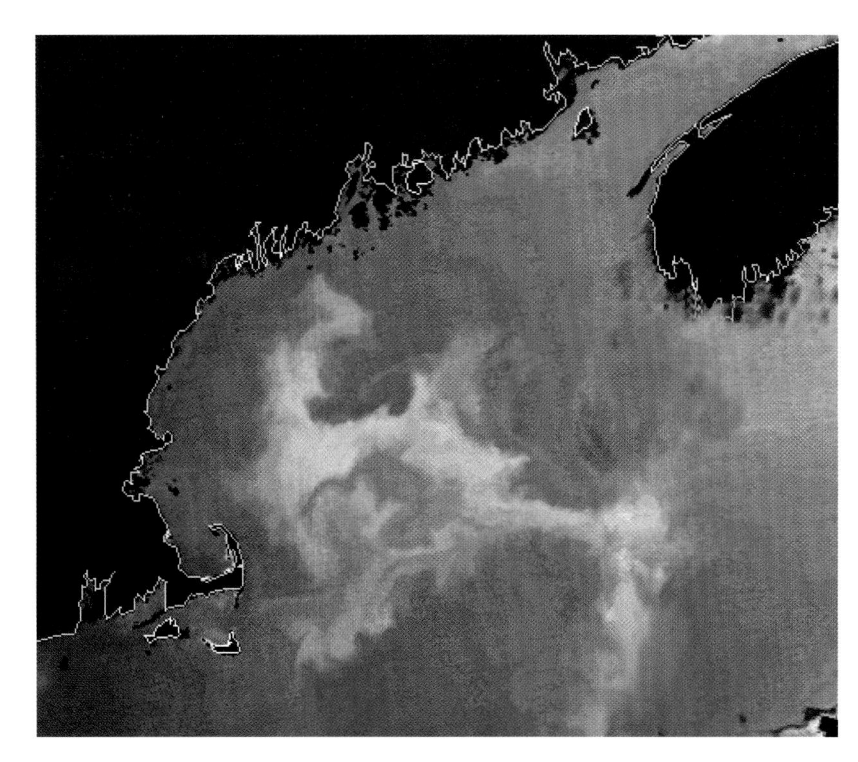

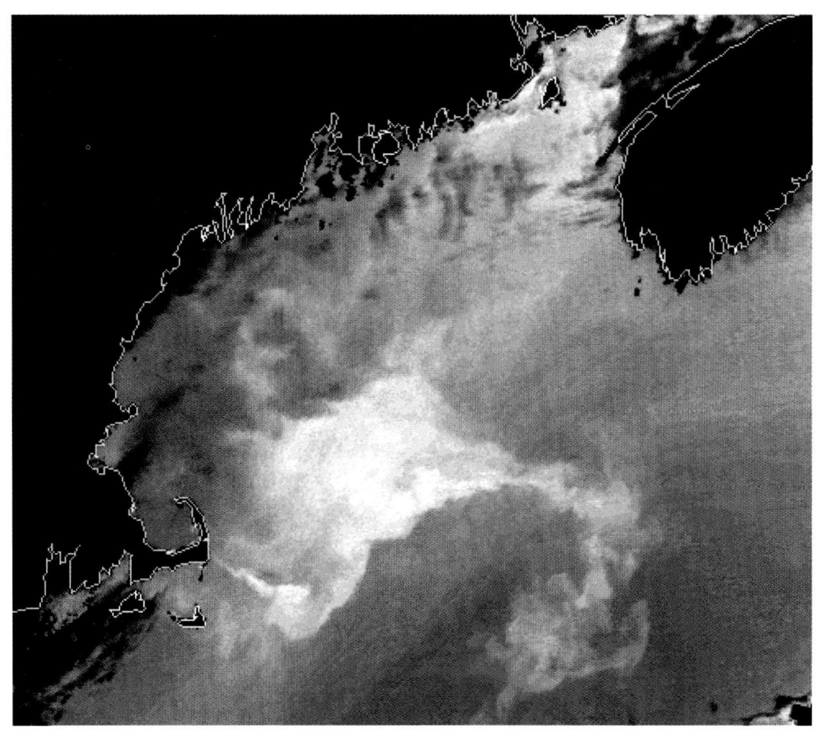

The reason for this functional youth lies in sea level changes and changes in absolute height of land that have occurred since and because of the great glaciation. The Gulf of Maine is a nearly landlocked body of water, and it was even more so 10,000 years ago when Georges Bank, rising above sea level, shut off most water exchange with the North Atlantic Ocean, the source of our tides. It was only about 4,000 years ago that melting glaciers raised sea level to the point that Georges was submerged, establishing a broad connection between the ocean and our Gulf and thus establishing the tidal regime we know today. Without that tidal action and its many consequences the Gulf was a much different body of water. Try to imagine the coast of Maine or the Bay of Fundy with tides of two to three feet instead of 10 to 50 feet.

Our exceptional tides, therefore, are a very recent feature in the context of geological time. Prior to their onset about 4,000 years ago, the intertidal zone was much smaller than it is now, and the abundance of life dependent upon that zone must have been notably different, probably with many fewer intertidal and subtidal seaweeds, and certainly with smaller mussel and soft-shell clam populations.

Productivity of Dual Systems

In summary, we believe it is the combined effects of topography and climate that have given rise to a highly productive system in the Gulf of Maine. Productivity is maintained by two physical systems, which we can describe as solar-lunar and solar-thermal. The solar-lunar system allows the shoal areas such as Georges Bank to undergo continuous mixing, which provides nutrient-rich water for luxuriant phytoplankton growth. The solar-thermal system provides sunlight for photosynthesis and heat for stability. Seasonally these processes stratify and destratify the deep-water masses of the central region of the Gulf of Maine. With each sequence they produce bursts and/or recessions in phytoplankton growth.

Which system is most important? The seasonality of production described as "solar-thermal" is characteristic of many high-latitude regions of the oceans. What makes the Gulf of Maine unique is that it is surrounded by banks and shoals and a narrowed geometry that can focus tidal energy and translate it into biological productivity.

Topography (the depth of banks and shoals) is extremely important to the productivity of the Gulf of Maine. If the banks were deeper, then the needed vertical mixing would not occur. If the banks were more shoal, they would begin to support attached algae, which would alter the type of fisheries we find here. If the entry channel were shallower, there would be less nutrient-rich water entering the central basins. If the basins were shallower, the Gulf of Maine would undoubtedly become an unproductive arm of the Atlantic Ocean.

Through the combined forces of geology and climate, a basin has emerged whose overall environment for the production of planktonic marine life seems ideal; a garden where climate, topographic geometry, and general oceanographic factors come together in productive harmony.

5.1 Inshore Herring Fishery

Brailing the pocket of a herring weir in Rockport Harbor, Maine.
Credit: photo by Harlan Hurd, courtesy of Carroll Thayer Berry Collection, Penobscot Marine Museum

5

A PECULIAR PIECE OF WATER

UNDERSTANDING FISH DISTRIBUTION IN THE GULF OF MAINE

Spencer Apollonio and Kenneth Mann

A nineteenth-century historian called the Gulf of Maine "a very marked and peculiar piece of water," and indeed it is. The Gulf is quite distinct from the Atlantic Ocean, differing by such significant measures as geological history, water characteristics, age, and productivity. Its marked and peculiar nature could be a great help in understanding the relationships of the dynamics of the sea with those of the plants and animals that live there, if we were wise enough to read the signs (fig. 5.1).

The fisheries of the Gulf support about 20,000 fishermen and each year produce about 530,000 metric tons of shellfish and finfish worth about $650 million. The well-being of these fisheries increasingly depends upon management that understands the physical and biological interactions within the system that supports them. The collapse of the Newfoundland cod fishery and the suspension of fishing for all groundfish in the Atlantic provinces of Canada has resulted in severe economic and social dislocations in hundreds of communities. These are tragic examples of the failure to think systematically.

Part of the reason it has been difficult to understand ecological relationships in places like the Gulf of Maine is that bodies of water are almost impossible to see. Oceanographers have collected tens of thousands of bottles of seawater, analyzed their temperature and salinity, and set countless more bottles and buoys adrift to chart currents—a labor-intensive process from which one can only infer the larger picture, not actually see it (fig. 5.2). Now, with the aid of satellite imagery, it is possible to display different layers of information in a single picture to gain a better understanding of how oceanography, geology, and marine biology are intertwined and how together they underlie the marine food chains on which all life in the Gulf—including ultimately our own—depends. Although such imagery cannot reveal important components in the Gulf of Maine, such as the bottom-water currents and many fish distributions, it is an increasingly useful tool for creating an integrated picture of the whole Gulf.

5.2 Collecting Seawater Samples
(above)

This method of collecting oceanographic samples is time-consuming and expensive. *Credit: courtesy Bigelow Laboratory for Ocean Sciences*

5.3 The Gulf of Maine Gyre
(right)

The lines over this satellite image show the counterclockwise circulation of surface waters during typical summer conditions in the Gulf. The gyre is influenced by the strong Scotian current, which sweeps northward past the shores of Nova Scotia. Off Nova Scotia the current divides, part continuing north and east into the Bay of Fundy while the rest swings westward over Grand Manan Banks and along the eastern Maine coast. Off Massachusetts the seaward edge of the current develops an easterly flow that gains in volume as the current approaches Cape Ann (near Gloucester). This easterly flow sweeps in a broad band across the southern half of the Gulf of Maine, skirting the northern edge of Georges Bank, where some of it rejoins the water from the Scotian Shelf to complete the gyre. The rest flows around the northeast peak of Georges Bank to set up a clockwise gyre over the center and edges of Georges. *Credit: image enhancement and graphics: S. Meyer, Island Institute, after data by H. Bigelow*

The Gulf of Maine Gyre

The surface waters of the Gulf of Maine flow counterclockwise in a great gyre, as discussed in chapter 3. Although the engine behind this flow comes in part from a cold coastal current off Nova Scotia that flows over Browns Bank and enters the Gulf of Maine near Cape Sable, long-term measurements indicate an additional flow of deeper waters from the Atlantic Ocean that enter the Gulf through the deep but narrow offshore cut, called Northeast Channel, between Georges and Browns banks (see fig. 4.4). Once these currents enter the Gulf, the earth's rotation deflects surface currents toward the shores of Nova Scotia and the land surrounding the Bay of Fundy, where they curve back around to the southwest (fig. 5.3). Off Casco Bay, the surface current trends more to the southeast and then to the east to set up a gentle clockwise gyre around Georges Bank.

Additional thrust to the circulation is supplied by spring runoff of fresh water from rivers, as well as from one of the world's most powerful tidal surges. It has been estimated that the outer edge of this anticyclonic (counterclockwise) gyre of water rotates at a rate of seven nautical miles a day. Thus it takes approximately three months for the surface waters of the Gulf to complete a single revolution. In contrast, the bottom waters of the Gulf cycle through the deep basins over a period of about one year.

In places, tidal currents are so strong that they mix the water from top to bottom and prevent the formation of distinct layers. These tidally mixed areas have several distinct characteristics. One is that the temperature at the surface tends to be cooler in summer and warmer in winter than elsewhere. Another is that plankton production is spread more evenly throughout the year. A third is that the tidal currents carry away the fine sediments, creating a gravel bottom much favored by herring and other species for spawning. As mentioned before, the top of Georges Bank is a tidally mixed area, and this is one of the factors accounting for its high productivity of plankton, shellfish,

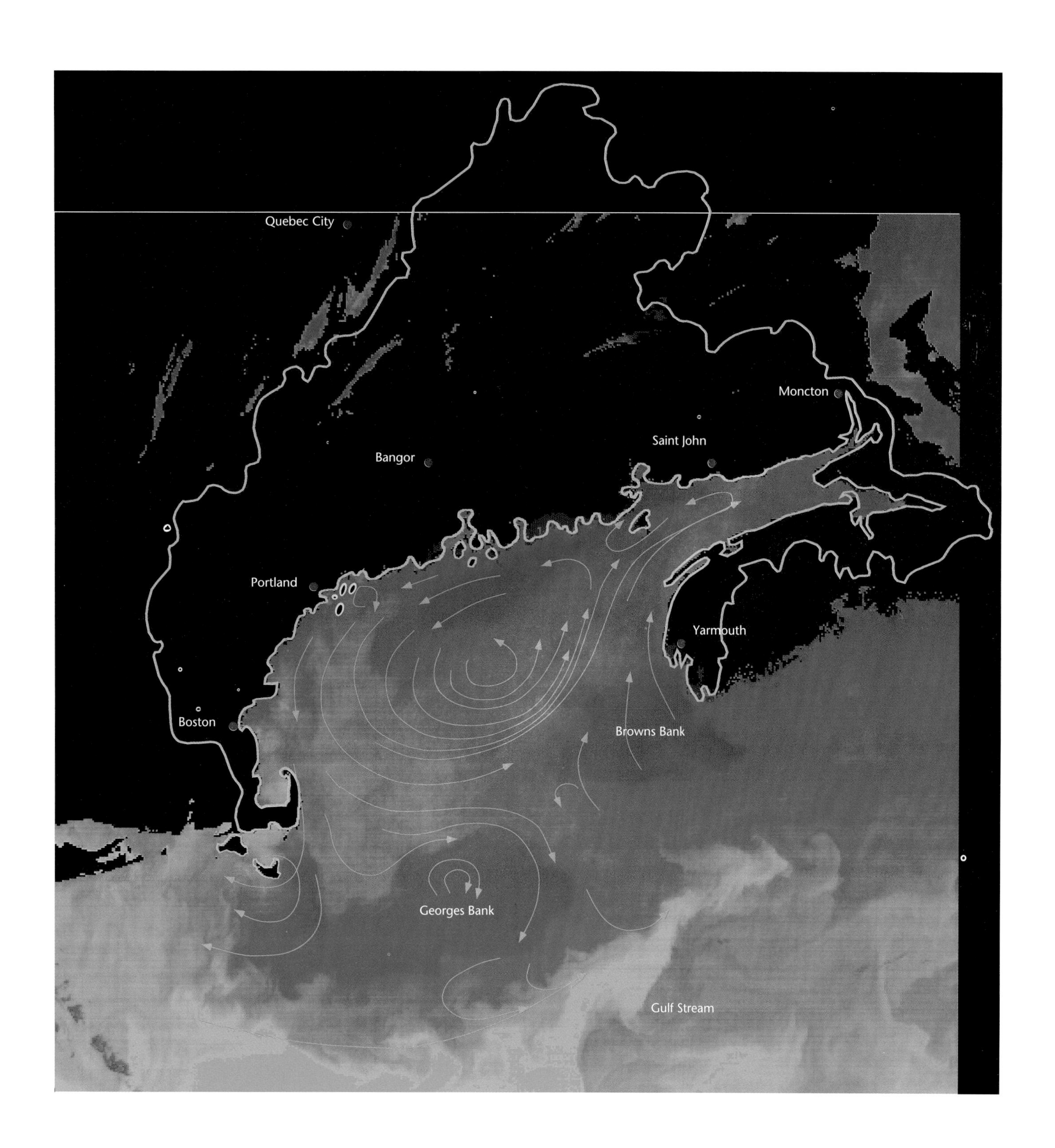
Quebec City
Moncton
Saint John
Bangor
Portland
Yarmouth
Boston
Browns Bank
Georges Bank
Gulf Stream

and fish. There are large areas in the northern Gulf that are tidally mixed, and their relatively warm winter temperatures make them suitable for salmon rearing. In summer, the dividing line between a tidally mixed area and a stratified area becomes a tidal front, and this is a site of strong biological activity.

Seawater Structures in the Gulf of Maine

The structure of different bodies of water in the Gulf, which we might call hydrographic structures, are just as real as its geographical units and have as much significance for the inhabitants of the Gulf. These include horizontal layerings of water caused by characteristic mixes of temperature and salt content, as well as *fronts* and *upwellings* that occur seasonally in some areas within the water column. For example, ocean surface temperatures off southwest Nova Scotia are cold even in summer (see fig. 4.6), indicating upwellings in that area: because currents move surface waters away from that coast, deeper, cold waters, rich in nutrients for the plankton, rise to the surface to replace them. And there we find a rich population of seabirds, attracted by the tiny floating planktonic animals whose growth is stimulated by that upwelling. Upwellings can also occur in inshore areas with strong tidal surges, as well as where currents force deep water over shallow offshore banks. One of the most remarkable upwellings, Old Sow whirlpool, can be seen from the Deer Island ferry in Passamaquoddy Bay as it runs between Eastport and Deer Island. Other upwellings are found in several areas at the mouth of the Bay of Fundy. These are places of exceptional biological activity because the currents stir bottom nutrients into the water column, making it available as food for marine species (fig. 5.4).

Another important hydrographic feature is that during most of the year the Gulf has a three-layered structure: surface water that varies in temperature and salinity with the seasons, an intermediate layer that is usually cooler, and a warm, salty bottom layer. At the base of the surface layer, a thin layer of water characterized by a sharp temperature differential may form in summer at the boundary of the surface layer and the intermediate layer. Called a *thermocline* by oceanographers, this narrow band of water is not a permanent feature throughout the year. In fall and winter, the thermocline breaks down and surface and intermediate waters of the Gulf of Maine mix, but heavier, saltier bottom waters remain isolated.

At the outer edge of the Gulf, where cold waters circulating along the seaward edge of Georges Bank meet the warmer waters of the Gulf Stream, a temperature front develops (fig. 5.5). Fronts are the boundaries of water bodies of different temperatures, densities, and salinities, analogous to weather fronts in air masses. Oceanographic fronts are places of exceptional biological activity. From the scientific and management points of view, it is significant that these fronts are all regular and predictable features of the Gulf. They figure prominently in the distributions, abundances, and life characteristics of a number of species of fish, marine mammals, and seabirds.

Temperature and Stratification

The distributions of some species within the Gulf of Maine appear to be directly related to the variations of sea surface temperatures. For example, surface temperatures off Cape Ann and over Jeffreys Ledge on the western side of the Gulf are warm in the summer; this is the warmest part of the Gulf, reflecting the strong temperature stratification that prevails here during the summer and early fall. These warm surface waters impede vertical mixing in the water column. They form a blanket that insulates the cold bottom waters from the warming effects of summer air temperatures (fig. 5.6). In fact, the bottom waters of this area in summer have the coldest temperatures of the Gulf at any season. Here we find northern shrimp in greatest abundance: shrimp, a subarctic animal, prefers these cold waters.

Along the Washington County coast in eastern Maine, in contrast, cool summer surface temperatures mark a well-

5.4 Vertical Mixing off Jonesport, Maine

Areas of vertical mixing occur where intense tidal activity stirs the entire water column, such as off Jonesport, Maine. These are places of exceptional biological activity because upwelling currents stir nutrients from the bottom into the water column, creating conditions conducive to phytoplankton blooms which feed shellfish and attract concentrations of fish. This image shows the eastern Maine coastal current, a tongue of cold, vertically mixed, nutrient-rich water that circulates at the northern edge of the Gulf of Maine gyre. *Credit: image enhancement: Island Institute, S. Meyer*

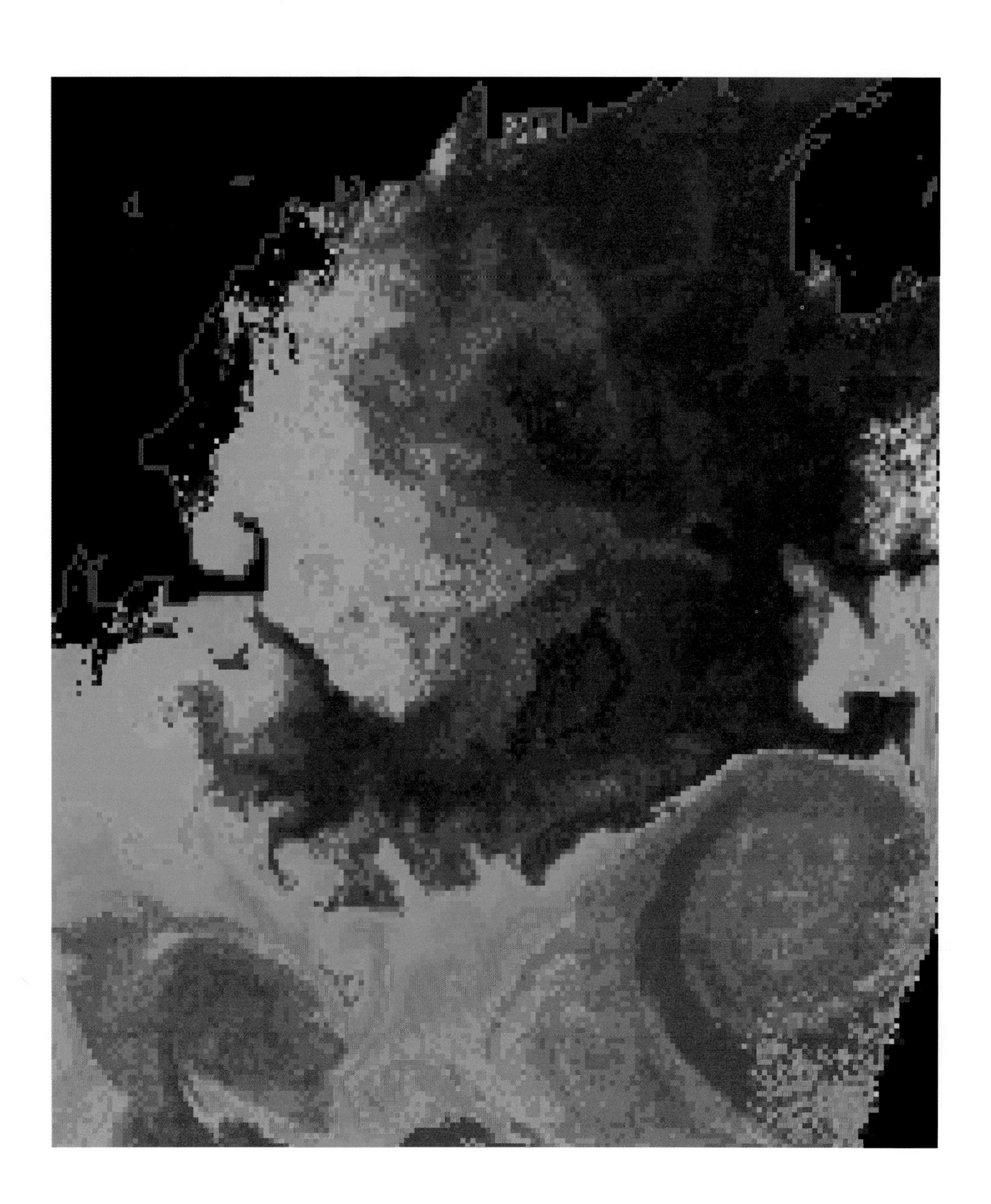

5.5 The Warm Core Ring
(left)

A warm core eddy, or ring (the orange-red circle), has spun off the Gulf Stream and is shown careening up against Georges Bank in this May 27, 1993, Advanced Very High Resolution Radiometer (AVHRR) image. This warm core ring registers an 8-degree temperature difference, dramatically different from the blue waters of the Gulf of Maine. This ring is at its peak and is likely to dissipate in 4 to 5 hours. The 200-foot drop off the southern edge of Georges Bank into the Atlantic is virtually a wall keeping the ring in place. The smaller eddy to the left will probably take its place. Whales are known to travel hundreds of miles to feed along the perimeters of these Gulf Stream eddies. The black area in the lower right is a cloud mass; the pink area to the right of Georges Bank is probably haze or light cirrus clouds. *Credit: Satlantic*

5.6 Cross Section of the Basins of the Gulf of Maine
(below)

Temperature of water masses, showing stratification over the deep basins of the Gulf of Maine in summer and winter. *Credit: T. Christensen; data: National Oceanographic Data Center files for the years 1940–1980*

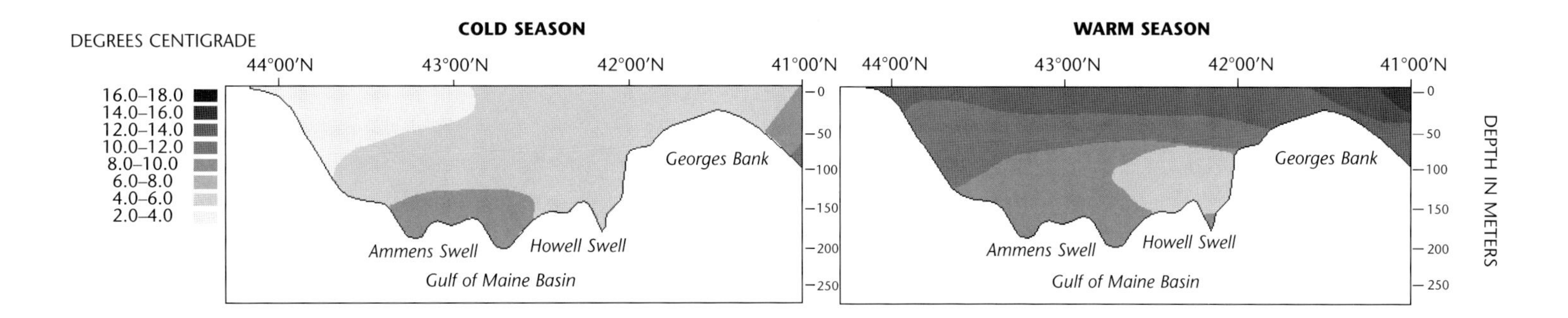

defined tongue of water that parallels the shore. These cool temperatures indicate considerable vertical mixing largely due to tides ebbing and flowing in the Bay of Fundy, the effect of which is to bring nutrients such as nitrates to the surface (fig. 5.7). Sun-warmed surface waters are stirred deep; as a result, the bottom waters here are relatively warm in summer.

The tidal mixing off the eastern coast of Maine means that surface waters do not cool down as much in winter as do the surface waters further west. These characteristic temperature patterns account for the fact that eastern Maine and southwest New Brunswick are commercially important salmon aquaculture areas (fig. 5.8). The relatively warm waters that come to the surface there in winter enable salmon to survive and grow in the cold months, while pen-rearing salmon is more problematic to the west.

In summary, the dynamics of the Gulf's various currents and countercurrents as they circulate through basins and over ledges, combined with the great tidal ranges and different temperature regimes of water bodies, determine the presence or absence of mixing within water layers of the Gulf. Changes in freshwater discharges, solar heating, and surface winds can also have significant effects all the way through the food web, notably on fish populations.

It is often said that fish population cycles are unpredictable, and some do indeed fluctuate greatly from year to year. But changes in fish populations are not well understood because scientists have focused on fish, rather than on the ecosystem.

Patterns of Fish Distribution

Some species of fish live in the Gulf year-round. They may have local movements with the seasons, but the Gulf is their home. Most of the cod family (including haddock and three species of hakes), ocean perch, and six species of flounders are such residents. But as we shall note, they have interesting distributions within the Gulf.

5.7 The Benthic Pump

High surface-level concentrations of nitrates (measured in micrograms per liter of seawater) coincide with the tongue of cold water (dark blue) that trends along the eastern part of the coast of Maine about 10 miles offshore. This is an area of important vertical water movement that brings much higher concentrations of nutrients such as nitrates to the surface. This band of nutrient-rich cold water stretches from the southern end of Grand Manan toward Schoodic Point and Matinicus Island in outer Penobscot Bay. *Credit: satellite image enhancement: S. Meyer, Island Institute, after data by S. Apollonio*

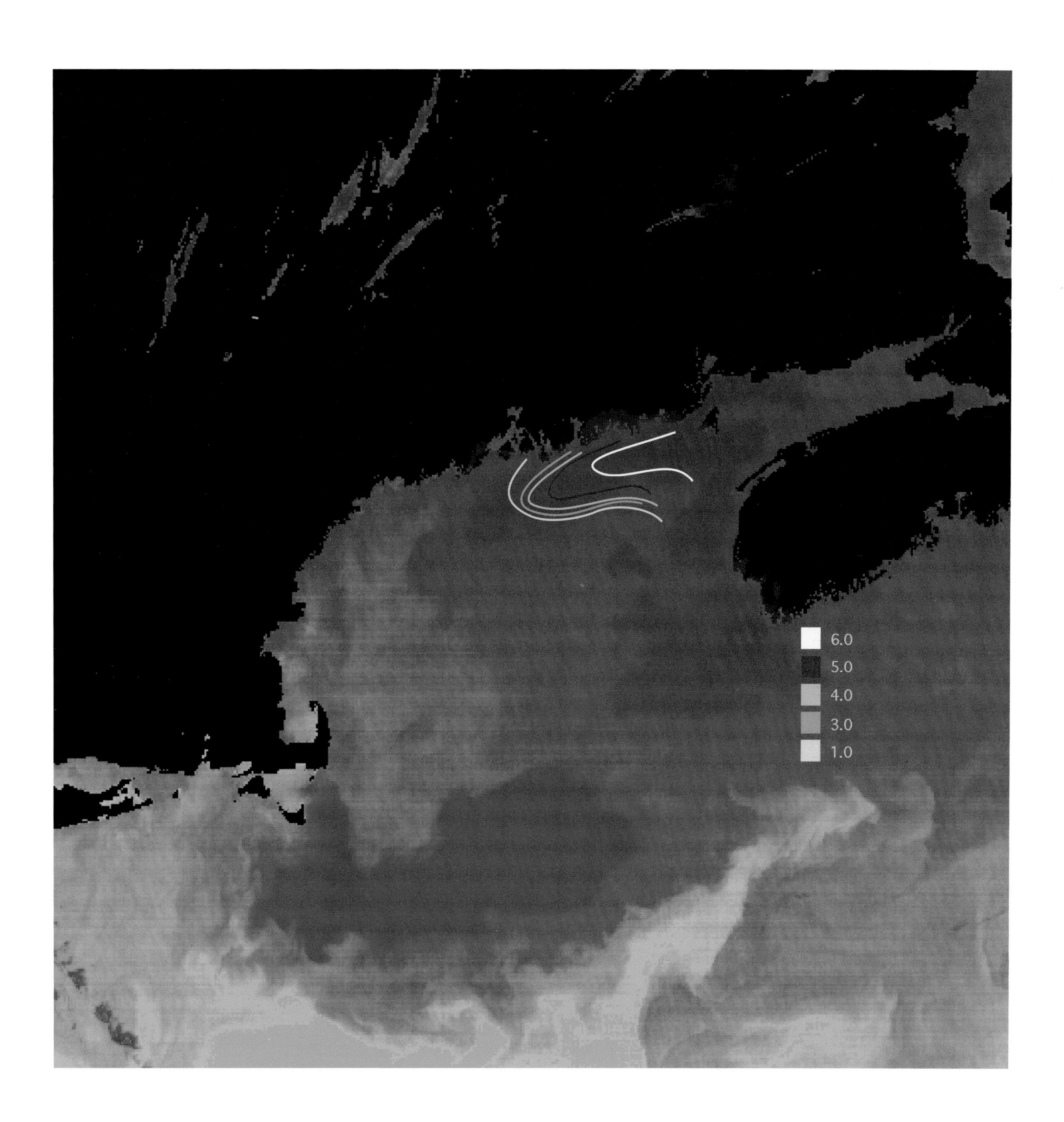
6.0
5.0
4.0
3.0
1.0

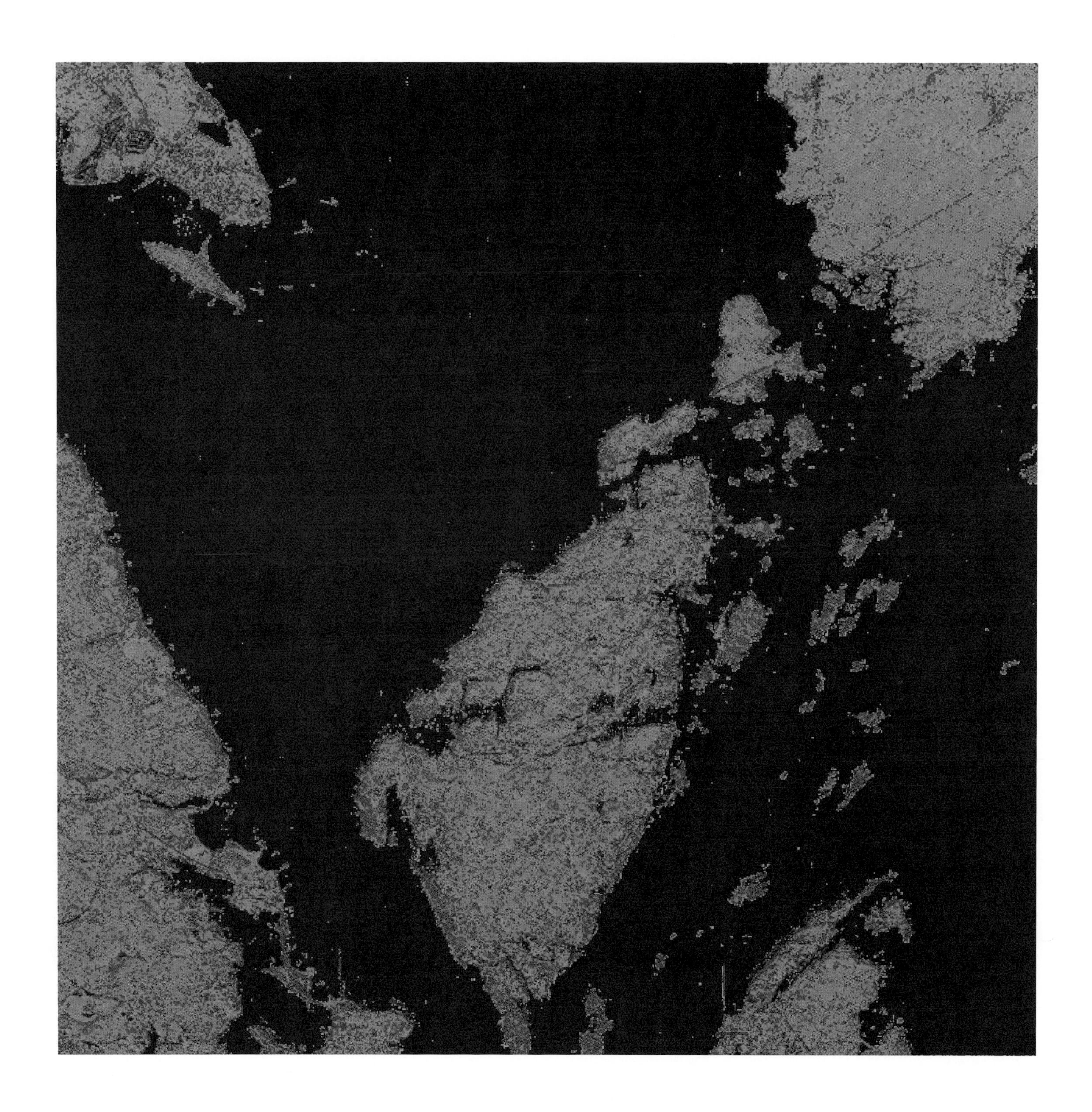

5.8 Passamaquoddy Bay Salmon Pens and Herring Weirs

Satellite image of herring weirs (V-shaped structures) and salmon pens (small circles) in Passamaquoddy Bay at the Maine–New Brunswick border. Using radar remote sensing, local authorities are able to estimate sizes of cages used in salmon farms and lobster ponds, and herring weirs as well, with 80 percent accuracy. This allows for efficient monitoring of aquaculture practices over a large area. Radar remote sensing has the advantage of being able to read through clouds, rain, and fog and can therefore acquire data on a given site at any time.

The data shown here were acquired July 6, 1989, by the Canadian Centre for Remote Sensing using an airborne SAR (Synthetic Aperture Radar) sensor. The spatial resolution is 6 meters. The case of the weirs is particularly interesting since they are identifiable by the radar not because of their size but because of their structure. The weirs consist of vertical wooden structures with a diameter of 30 centimeters spaced approximately 1 meter apart. This light construction is less than what the radar can detect. The physical phenomenon that permits their identification is called corner reflection. The radar signal is reflected almost entirely within the dihedron formed by the plane containing the wooden structures and the plane of the water surface, creating a nearly perfect image of the weirs. Depending on the nature of the material (e.g., metal, wood, plastic), these sites tend to provide high radar returns, thus allowing identification and location on radar imagery.

Due to Canadian government support, salmon aquaculture became a viable business within a decade, revenues growing fivefold from the early 1980s to early 1990s. This rapid development created some problems. Ideal sites for establishing an aquaculture industry, for instance in the Passamaquoddy Bay and the Grand Manan area, have long been used by the traditional fisheries. Conflict between local fishermen and aquaculture farmers has focused on competition for space and concerns over biological interference and environmental impacts. *Credit: image enhancement: T. Ongaro, Island Institute; data courtesy of the Canadian Centre for Remote Sensing*

Bluefin tuna are spectacular migrants through the Gulf, which is simply a way station in the journeys that take them over much of the North Atlantic. More modest but nevertheless impressive migrants are mackerel moving from deep waters off the Virginia capes to the Gulf of Maine and the Gulf of St. Lawrence. Atlantic salmon, sturgeon, sharks, menhaden, alewives, some herrings, shad, dogfish, and squids all move in and out of the Gulf of Maine seasonally, just as do all the whales and most of the other marine mammals. One must be impressed by the mass and precision of these great seasonal movements, each species moving purposefully along pathways clearly marked to them by signals hardly suspected by us.

The only fish migrants coming to the Gulf from farther north are the Atlantic salmon returning home and the occasional capelin, a subarctic member of the smelt family. The rest of our migrants, whether fish or mammals, come from the south to feed and spawn in the rich waters of the Gulf during the summer and fall, and cycle out of the Gulf in the winter.

Fishermen know that fish are not scattered randomly about the Gulf; one must look for certain species in certain places at certain seasons. These patterns of distribution are more than geographical curiosities; they appear to be related to the fundamental ecological structure of the waters of the Gulf and the biology of the species. By taking fully into account these patterns, we may gain a much better knowledge of how fish species behave under various conditions.

There is a working hypothesis that the distributions of *pelagic* (near-surface) fishes, like herrings, and of benthic or *demersal* (bottom-living) fishes, like cods and flounders, in the Gulf of Maine are direct reflections of the differing regimes of mixing along the coast of Maine. The demersal fishes seem to be more abundant along the western parts of the coast, while herrings predominate along the eastern parts of the coast and off southwestern New Brunswick. These differences appear to coincide, respectively, with the more stable water in the west and with the mixed waters in the east.

In the west, according to this hypothesis, the basic phytoplankton food supply is produced in a strong pulse each spring and then sinks to the bottom, where it supports the benthic animals upon which demersal fish feed. In the east, in contrast, phytoplankton productivity occurs more evenly throughout the year and tends to remain in the water column rather than sinking to the bottom, because of the strong tidal mixing. These two factors favor the development of the pelagic, surface-feeding herring population.

The hypothesis remains to be proven, but it does suggest the important interaction, not sufficiently taken into account in fisheries management, of fish with their environment.

Herring and Mackerel

The abundances of mackerel and herring in western and eastern Maine respectively are familiar to fishermen. These distributions may reflect fundamental differences in these seemingly similar surface-schooling fish. Herring are dependent for their spawning on tidally mixed areas bounded by oceanographic fronts, which in summer separate cooler from warmer waters and in winter separate the tidally mixed areas from the wind-mixed areas. Fronts are predictable and quite permanent features of Gulf of Maine waters because they are functions of water depths and tidal currents (and thus of the moon). Herring, like other animals with their kind of life history, have come to rely on these physical features of the ocean for their basic reproductive patterns (fig. 5.9).

Mackerel reflect a different kind of life history; they do not search out particular waters, but instead shed their eggs, apparently unpredictably, wherever may be convenient during the course of their extensive migrations along the coast from the deep waters off the Virginia capes to the Gulf of Maine or to the Gulf of St. Lawrence. These differences suggest basic differences in the vulnerability of these species to natural variabilities. They should also suggest basic differences in the management tactics that may be appropriate for each species.

Shrimp

Perhaps the best-documented example of the relationship between fish population and the physical structures of the Gulf is found in the shrimp population. Shrimp distribution coincides, we have noted, with the greatest temperature stratification of the Gulf, as well as with the coldest bottom temperatures at any season. Because of this stratification, shrimp are found in greatest abundance in the southwestern part of the Gulf of Maine.

But shrimp also swim upward and downward each night, passing through different parts of that temperature stratification in the water column, and researchers have learned that their egg production is affected by those temperature changes. The greater the temperature differences, the greater the egg production. Presumably the differences in egg production from year to year have some bearing on the subsequent abundance of shrimp catches by fishermen. Shrimp decline drastically in warm temperatures, but they also increase dramatically in cooler waters. It appears possible to predict future relative abundances simply by measuring the vertical temperature distributions in early summer when eggs are being formed.

Redfish

Redfish, or ocean perch, generally are found in greatest abundance in the deep basins, such as Truxton, Jordan, or Crowell, of the central Gulf. It is in those deeps that relatively warm waters and the waters with least temperature changes are found throughout the year.

Redfish are attuned to the rather stable environment they live in. They are slow-growing and reproduce late in life, producing a remarkably small number of young. Their biology makes them highly vulnerable to heavy fishing pressure, because

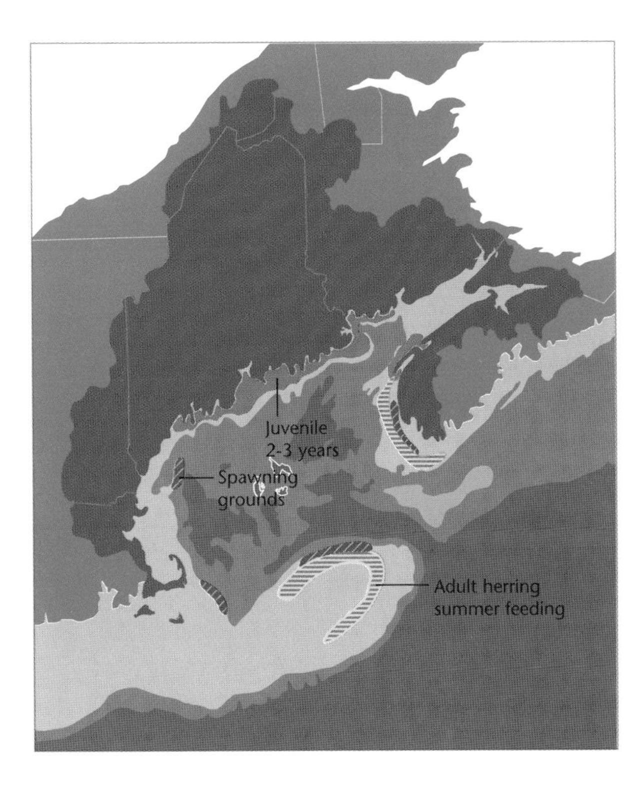

5.9 Herring Spawning Grounds

The distribution of adult herring on different spawning grounds in the Gulf of Maine shows a distinctive cycle and seasonality. In the spring in the upper Bay of Fundy, herring populations begin spawning in May. By June the herring grounds off Grand Manan and Passamaquoddy Bay see concentrations of spawning herring. Off the eastern coast of Maine, herring spawn in August; off central and western Maine, in September; in the western portions of the Gulf (including Jeffreys Ledge) and on Nantucket Shoals, in October; and on the rim of Georges and Browns banks, in November. *Credit: graphic by S. Meyer and P. Conkling, after data by Goode, Halliday, D. Stevensen, R. Stephenson, and Chang*

they are unable to recover quickly from stock reductions (either from heavy fishing or from natural catastrophes). Redfish used to be an important commercial fishery in the Gulf of Maine; it is no longer. We failed to understand their vulnerability to fishing pressure, even though that vulnerability is obvious in their biology.

The Cod Family

Intriguing patterns of distribution are found among the great family of cod and cod-related fishes, commercially the most important family of fish in the Northern Hemisphere. Found primarily on the continental shelves in temperate and cold water, cod, haddock, and hake used to account for almost one-fifth of the world's fish catch. Today, due to severe fishing pressure, the catch has fallen to 10 percent of the world's total.

The cod family, or gadids, includes cod, haddock, pollack, cusk, whiting or silver hake, and red and white hakes. Each has its pattern of distribution throughout the Gulf of Maine. From the distributions of cod, haddock, and pollack, it is clear that each species shows a kind of alternation, or reciprocity, of relative abundance within particular areas (figs. 5.10, 5.11, 5.12, 5.13). For example, whereas haddock tend to be more abundant on the eastern end of Georges Bank, cod tend to be more abundant in the nearshore waters of the Gulf.

The distributions of the cod family have a number of potentially important implications. With each species able to function best in a certain part of the Gulf, it is as if nature has hedged its bet, so to speak, to insure that whatever environmental disturbances there may be, there is always a species of this family to fill the gadid niche within the system. The spawning of gadid species also has a pattern in time; there seems to be a cod of one kind or another spawning throughout the year. We might thus say that gadids are important in some way that we do not fully understand for the functioning of the system, but that nature, perhaps, doesn't care whether a cod or a haddock or

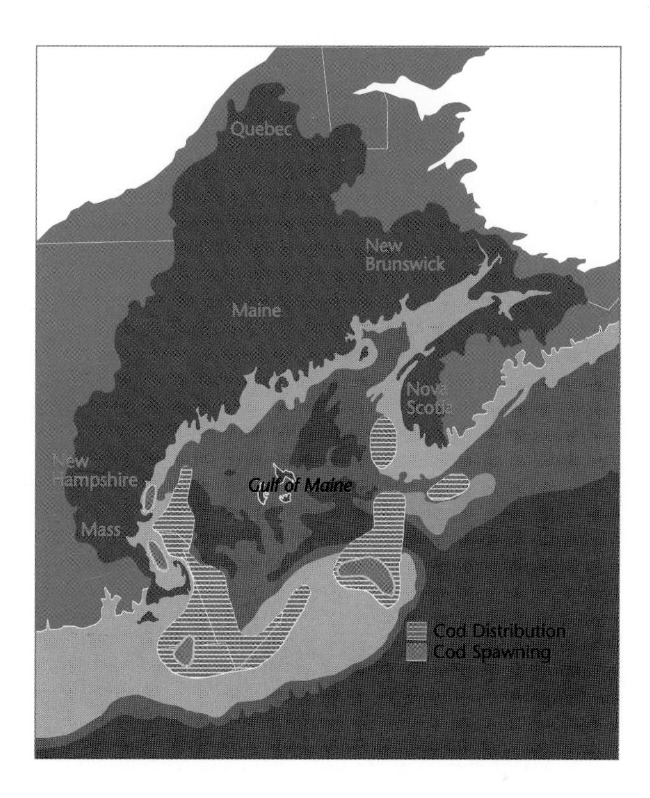

5.10 Cod Distribution

Distribution of primary concentration and spawning grounds for cod in the Gulf of Maine and on Georges Bank. Note the large spawning areas south of Nantucket and along the Massachusetts coast, especially between Cape Ann and Isles of Shoals, New Hampshire. Smaller spawning grounds in the United States are also shown off Portland, in outer Sheepscot Bay, and around the outer islands of Penobscot Bay. In Canada the largest spawning ground is on the Northeast Peak of Georges and the tip of Browns Bank. *Credit: graphic: S. Meyer, Island Institute, after data by R. Edwards et al.*

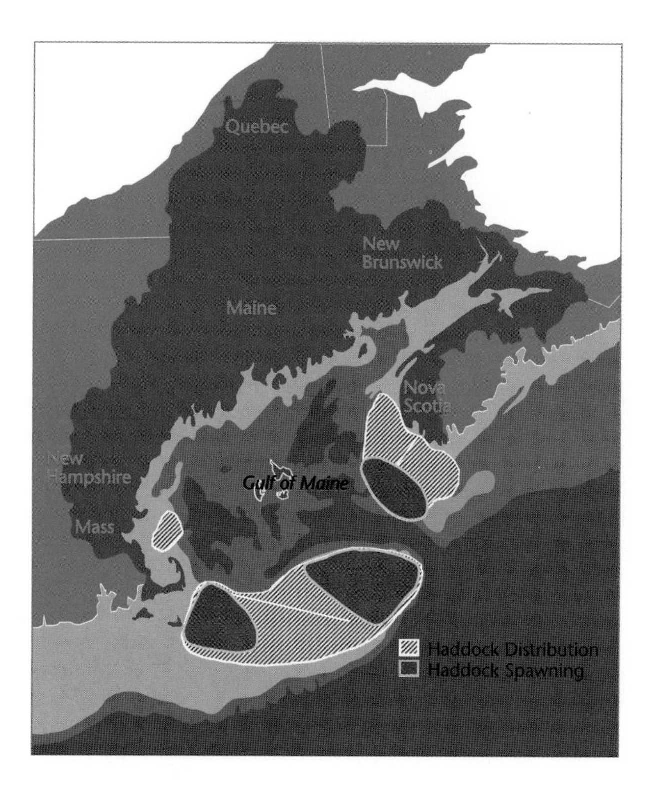

5.11 Haddock Distribution

The largest concentrations of haddock and most important spawning grounds are found on Georges Bank near Great South Channel and Northeast Peak and on Browns Bank. *Credit: graphic: S. Meyer, Island Institute, after data by R. Edwards et al.*

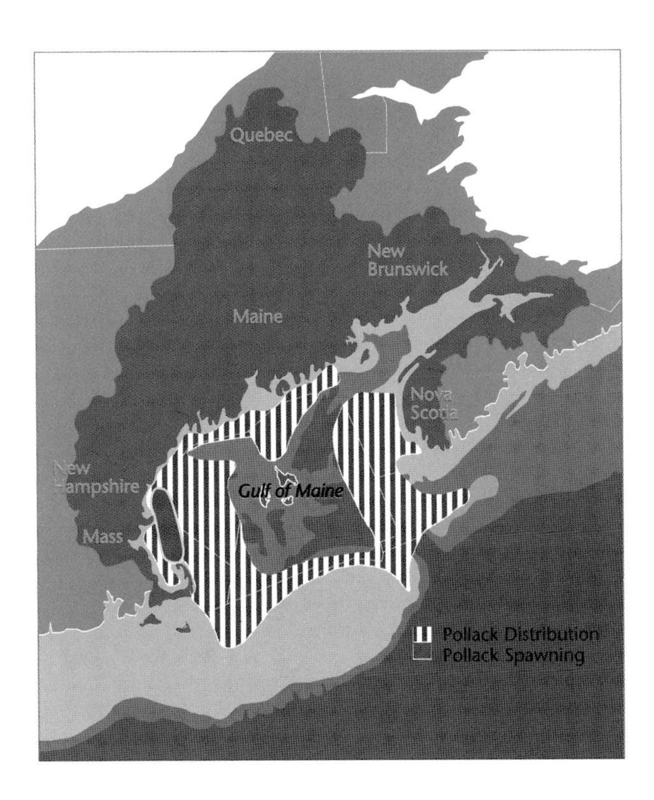

5.12 Pollack Distribution

Pollack is more widely distributed than either of its close cousins, cod or haddock. It is found in commercial concentrations in deeper waters that rim the Nova Scotia and Maine coasts. *Credit: graphic: S. Meyer, Island Institute, after data by R. Edwards et al.*

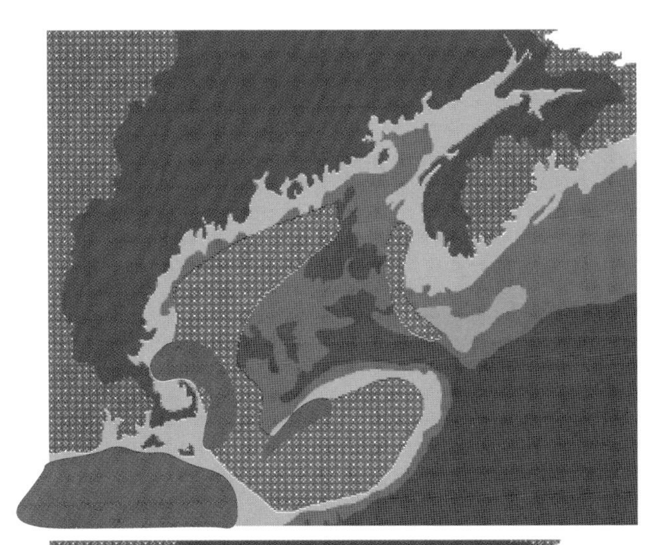

5.13 Red and White Hake Distribution

Resource distribution of red hake (top) and white hake (bottom) in the Gulf of Maine area. Shaded areas (in pink and red) show major spawning grounds of the two species. Inshore waters of the coast of Maine from Eastport to Mount Desert Island are important spawning grounds for white hake, while red hake spawn in the western portion of the Gulf of Maine and on the northern rim Georges Bank. Stocks of red hake in the inner Gulf of Maine appear to be separate from those on Georges Bank. *Credit: graphic: S. Meyer, Island Institute, after data by R. Edwards et al.*

a cusk performs the function. The important thing these distributions seem to imply is that the Gulf of Maine ecosystem has evolved with, and may depend upon, many species of the cod family being present in abundance.

The patchiness of the distributions of each species further reinforces our suspicion that the patterns of abundance reflect a mechanism to stabilize the system against the vagaries of the environment. Patchiness, or spatial heterogeneity of distribution, is a way of absorbing the perturbations of nature so that disturbances are confined to a small population that receives the immediate impact. The rest of the system is insulated from that impact and remains undisturbed. It is interesting to note that this phenomenon is found in populations like cod and haddock that show little migration or intermixing between semidistinct stocks (fig. 5.14). Patchiness within mobile fish groups, like mackerel, would have little adaptive benefit.

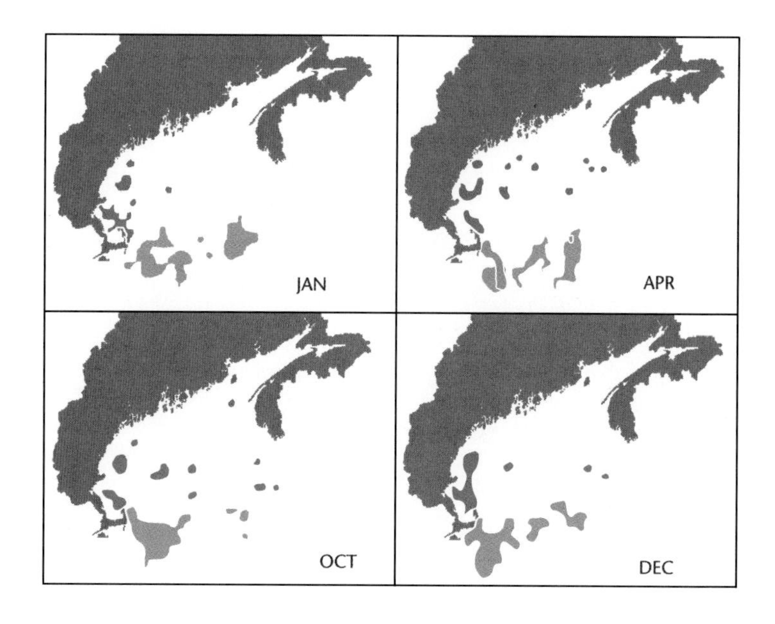

5.14 Cod Patchiness

In the course of the year, major concentrations of cod shift around Georges Banks and the Gulf of Maine, as shown in these graphics from January, April, July, and December. *Credit: S. Meyer, Island Institute, after data by Halliday*

The Collapse of the Northern Groundfish

In the summer of 1993, Canada's Department of Fisheries and Oceans (DFO) released information on groundfish stocks in Atlantic Canada that sent shock waves through Canadian communities. They reported the lowest number of fish ever recorded, the lowest-ever spawning stock of mature fish able to reproduce, slow growth and therefore smaller fish, and a very high percentage of fish being killed by the fishery, predators, and ocean conditions; and they predicted continuing declines. Many stocks are on the verge of commercial extinction. Only dramatic reductions in catches, if not total shutdowns of some fisheries, will prevent ultimate collapse.

The fisheries of the Gulf of Maine in Canada and New England, like fisheries elsewhere in the world, are managed on a species-by-species basis. Fisheries scientists and managers try to calculate numerical levels for maximum sustainable yield for cod, flounder, and so forth, and then regulate fishermen through quotas or other means based on these numbers. But throughout the major fisheries of the world, of which the northern groundfishery is only one example, these management approaches have failed to prevent overfishing (fig. 5.15).

In Atlantic Canada, where between 20,000 and 40,000 fishermen are out of work due to the government's shutdown of its groundfisheries, the tragedy is one of the starkest in any of the world's oceans. The sorry state of Canadian fisheries, as of those of the United States, is a result of decades of increased fishing power and efficiency in a large fleet of draggers equipped with ever more sophisticated technology.

These vessels tow or drag otter trawl nets over the bottom to catch groundfish, a technique invented in 1905 in the North Sea and brought across the North Atlantic shortly thereafter. The nets of the otter trawls are supposed to catch big fish and let the smaller fish slip between the mesh, with the hope that these fish will spawn before they are caught again. But fishermen and scientists have known for decades that otter trawling is not selective. Non-target species or undersized fish may make up from two-thirds to three-quarters of the total volume of fish brought to ship (fig. 5.16).

Populations of important marine fishes have crashed in other parts of the world, including Pacific sardine and anchovy off the west coast of Peru and Atlantic herring off the coast of England. In each of these cases, there are sufficient data to demonstrate that heavy fishing pressure coincided with a set of adverse environmental conditions, including changes in water temperature, the combined force apparently driving these fisheries to commercial extinction. In the case of the northern cod off Newfoundland in Canada, where a total fishing moratorium was imposed in 1992, it appears that changes in the intensity of winter winds over Labrador resulted in harsh winter conditions and greater ice accumulation, which, when melted in spring, created a larger volume of fresher, colder water. This water traveled to the Grand Banks in spring and summer and created unfavorable conditions for the reproduction of cod.

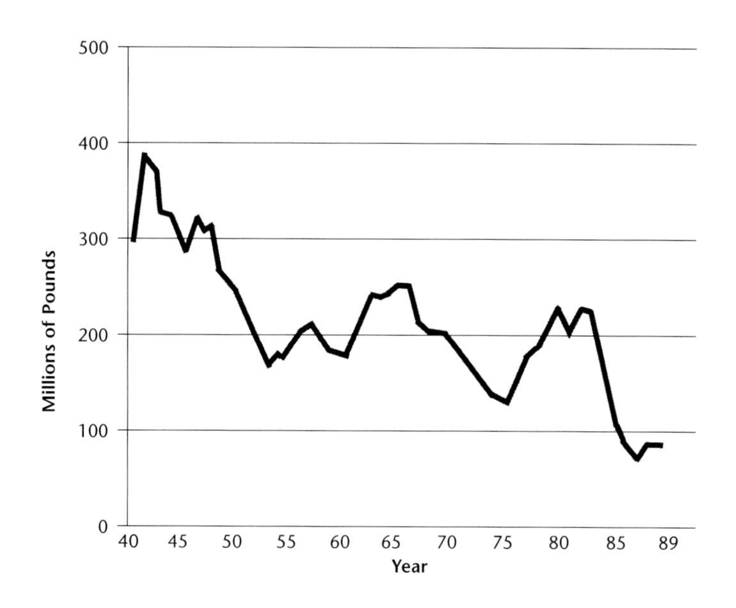

5.15 Groundfish Collapse

Domestic commercial landings of four major New England groundfish—cod, haddock, yellowtail flounder, and redfish—from 1940 to 1989, illustrating their decline. *Credit: T. Christensen*

5.16 Fishing Pressure

The increase in the size and power of fishing vessels in the Gulf of Maine, along with sophisticated electronic equipment, has contributed to the serious decline of most groundfish stocks. A catch of cod, haddock, and flounder is shown here in the process of being sorted on the deck of a fishing vessel out of Rockland, Maine, after the cod end of an otter trawl has been opened following a three-hour tow. Two-thirds to three-quarters of the total volume may consist of non-target species and undersized individuals, which must be tossed back overboard. *Credit: N. Parent*

As fisheries on both sides of the international border are being mismanaged, hundreds of millions of dollars and thousands of jobs of fishermen and other citizens in the region are lost. The problem is complex, but the roots lie in our failure to envision the fisheries of the Gulf of Maine and Georges Bank as interdependent marine ecosystems rather than as a collection of individual fisheries to divide up among those who vie for them. This failure is profound and systemic. If the fisheries are to be rebuilt into the powerful, sustaining source of wealth they once were, nothing short of a new scientific paradigm is needed to guide management decisions.

The Gulf of Maine as an Ecosystem

In the debates over how many fish of which species can be safely harvested, marine ecosystems are frequently referred to in a vague sort of way, but seldom has a "marine ecosystem" been defined. The absence of definition is at the heart of the problem. Essential to the concept of the ecosystem, formally defined, is that the whole system contains information that cannot be deduced purely from inspection of the parts. In the Gulf of Maine and on Georges Bank there are 52 commercially harvested species of fish and shellfish. Although vast amounts of data are gathered at great effort and expense, fisheries management plans on both sides of the international border are prepared species by species. Until we develop ecological models that relate the harvest of individual species to the marine system as a whole, management results will continue to be disappointing at best or destabilizing at worst.

Our Gulf is clearly not an accidental catchall of living things randomly scattered about in a homogeneous body of water. Scientists find instead much structure and much pattern in the Gulf and its inhabitants. We have hardly begun to describe these structures and patterns, even less to consider their significance; but clearly there are causal relations, with significance for the well-being and management of the abundance of life that we find here (fig. 5.17).

Ecosystems are not functionally described by geographical limits or by somewhat arbitrary species lists, even less by monitoring every physical and chemical variable that comes to mind—an impossible task even if it were useful. The concept of the ecosystem probably turns fundamentally upon the rates of such processes as energy flow and nutrient recycling, and upon the different kinds of species within the system. However an ecosystem is defined, it is clear that structure and pattern such as we find so clearly in both the oceanography and biology of this "very marked and peculiar piece of water" will be key parts of our understanding.

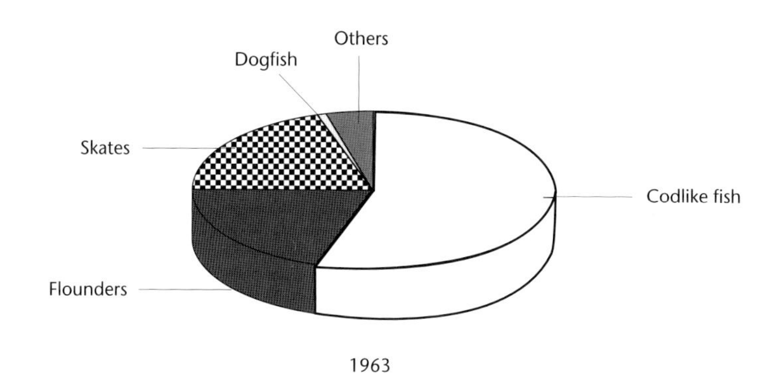

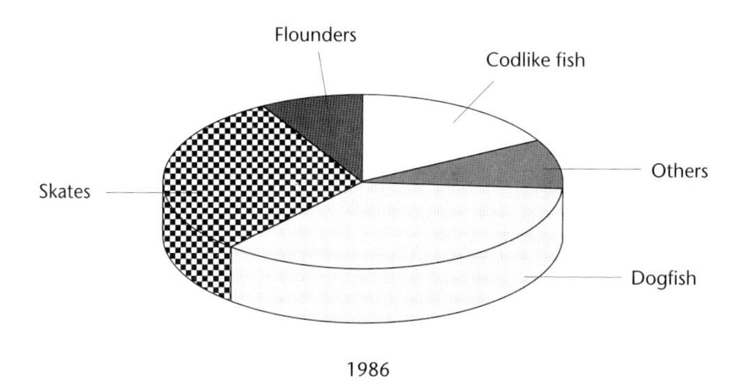

5.17 From Cod to Dogfish

According to National Marine Fisheries Service data, the proportion of codlike fish declined from 55 percent of the marine system in 1963 to 11 percent in 1986, while dogfish and skates, serious predators on cod and haddock, now comprise 74 percent of the biomass. Fisheries scientists do not know how serious or irreversible these changes may be. *Credit: data: National Marine Fisheries Service*

6.1 Atlantic Puffin

Atlantic puffins are at the south end of their breeding range in the Gulf of Maine. *Credit: P. Ralston*

6

THE TOP OF THE FOOD CHAIN

MARINE MAMMALS AND BIRDS OF THE GULF OF MAINE

David D. Platt, Richard Podolsky, Harry Thurston, and Janice Harvey

General Patterns of Distribution and Abundance

Marine mammals and birds share the resources of the Gulf's waters under conditions far different from their land-based counterparts. Marine mammals spend the bulk of their time gliding through a three-dimensional world; seabirds cruise over the two-dimensional surface of the sea. Rather than moving from one local habitat patch to another, whales and porpoises travel large distances from one region of the Gulf to another. Many species spend only part of their time here. Seabirds congregate along fronts at the boundaries between different types of water masses.

From the deck of a ship, the surface of the Gulf may look homogeneous, without much to offer in the way of ecological diversity. When viewed from space, however, the Gulf is surprisingly diverse and appears as patchy and heterogeneous as does the land from the same vantage point. Depending on the temperature, and especially on the amount and type of plankton suspended in the water column, seawater can take on a wide range of colors and properties. Marine mammals and seabirds key in on different water masses as they attempt to locate places that offer the best feeding opportunities.

Some areas of high marine productivity remain constant from day to day, even year to year. Others move from one location to another. Fishermen refer to portions of the sea that are predictably rich by the evocative term "ground." Rich fishing grounds in the Gulf of Maine are located above the shallow banks where currents rise up from deeper water, bringing nutrients. The upwelling of cold, nutrient-rich water stimulates biological activity and creates productive feeding grounds. Fishing grounds are also found at the mouths of large estuaries, where enrichment from drainage from inshore watersheds creates local areas of high productivity. Not surprisingly, most major seabird colonies and gathering places ("haulouts") of seals in the Gulf of Maine are located close to one or another of these features. Machias Seal Island, south of Grand Manan at the mouth of the

Bay of Fundy, for example, has an impressive concentration of seabirds and marine mammals dependent upon the productivity of an area where bottom waters come to the surface.

During the early years of New England whaling, rich grounds in the Gulf of Maine, such as Stellwagen Bank off Cape Cod, were visited by whalers regularly because whales could predictably be found there at certain times of the year. Today these same places are visited by dozens of whale-watching cruise boats carrying thousands of passengers each year. Right whales, reduced to a few hundred in the Atlantic by two centuries of aggressive whaling, breed in the Grand Manan basin at the Gulf of Maine's northern end. While all right whale areas are important on the Atlantic coast, this Canadian connection is critical to the species' survival because it is their only known nursery area, frequented by mothers and calves. The Roseway Basin, south of Cape Sable Island at the eastern edge of the Gulf of Maine, is their only known mating zone. The commercial whaling industry is a thing of the past in the Gulf, but whale watching (a multimillion dollar industry as well) likewise thrives or perishes on predicting where the whales will be.

Predicting where marine mammals and seabirds will concentrate is as simple as predicting where the richest waters are. For the great fishing and whale-watching grounds this is not too difficult. Locating the equally important but more ephemeral factors—such as warm water cores or eddies, or a temperature front between slope water and seawater circulating on Georges Bank—is more challenging. Images such as those provided by temperature-sensing satellites can be of great assistance.

Whales

It is striking that the five species of great whales that inhabit the Gulf—the humpback, fin, minke, sei, and right whales—are all baleen whales, adapted to grazing on the Gulf's immense quantity of plankton. As with many species of groundfish discussed

6.2 Right Whale Feeding Grounds

An interdisciplinary team of scientists from the University of Rhode Island, University of Washington, and Woods Hole Oceanographic Institution conducted a project called SCOPEX—the South Channel Ocean Productivity Experiment—to investigate the oceanography of the spring right whale feeding grounds in the Great South Channel region east of Cape Cod and Nantucket (visible on the left side of each image). Right whales are the world's most endangered large whale species, with only about 300 surviving off the east coast of the United States and Canada. These four images show the distribution of springtime right whale sightings overlaid on AVHRR-derived sea surface temperature patterns during mid to late May in four consecutive years, 1986 through 1989. Right whales are found in the central part of the region, north of the 100-meter depth contour (the V-shaped white line). South of the 100-meter contour, water depths are as shallow as 10 meters or less over Nantucket Shoals and Georges Bank; strong tidal currents keep the water column completely mixed year-round, with colder surface temperatures (darker blues). North of this contour, water depths increase to over 200 meters. Beginning in the spring, surface waters warm (lighter blues and greens) and the water column becomes strongly stratified. The mixed and stratified waters are separated by a tidal mixing front, which can be clearly seen on the AVHRR images as a boundary between cooler and warmer surface waters roughly parallel to the 100-meter contour. Right whales are consistently found in the stratified water mass and within a few kilometers of the tidal mixing front, where current patterns help to concentrate the zooplankton on which they feed into extremely dense patches. The whales tend to occur on the west side of the area in warmer years and on the east side in cooler years, due in part to slower development and growth by the zooplankton in colder waters. *Credit: R. Kenney, Graduate School of Oceanography, University of Rhode Island*

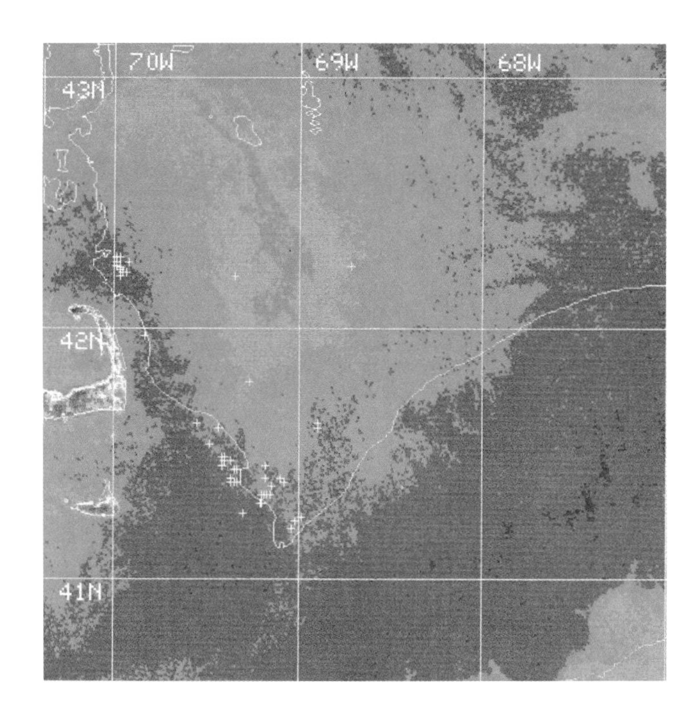
70W
69W
68W
43N
42N
41N

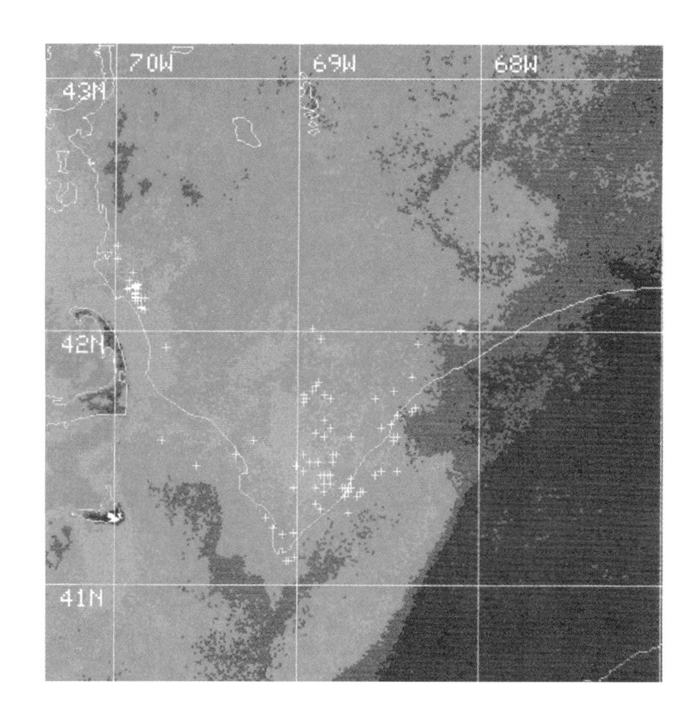
70W
69W
68W
43N
42N
41N

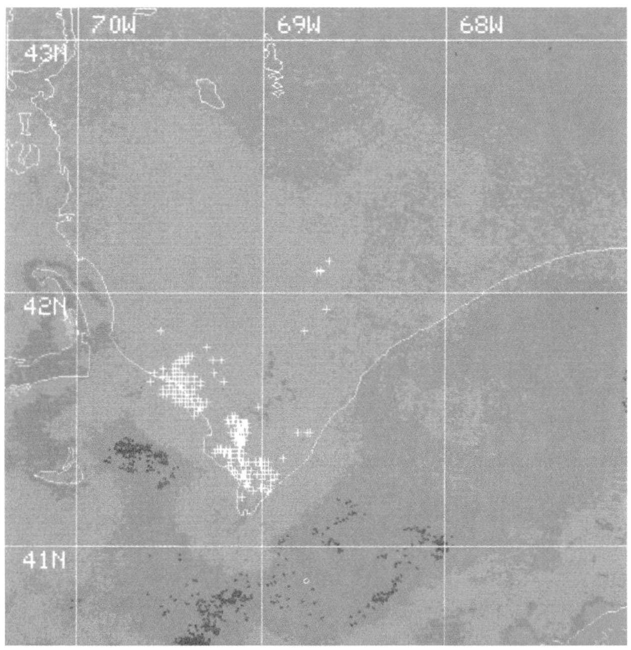
70W
69W
68W
43N
42N
41N

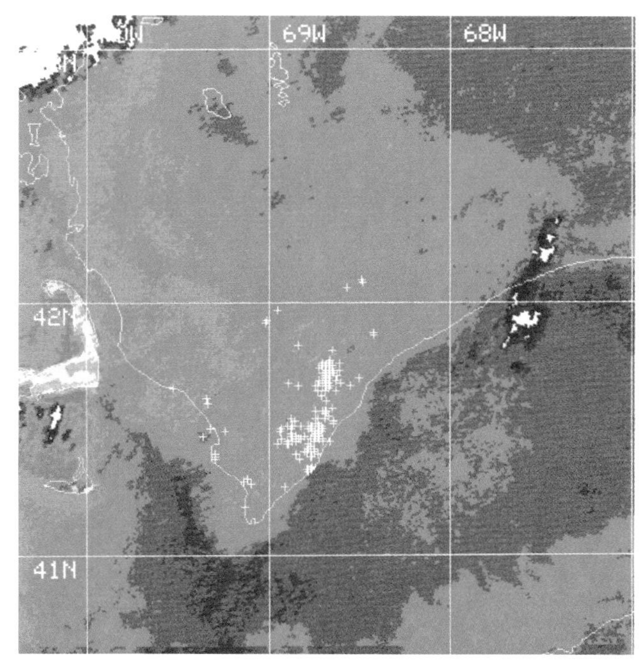
69W
68W
42N
41N

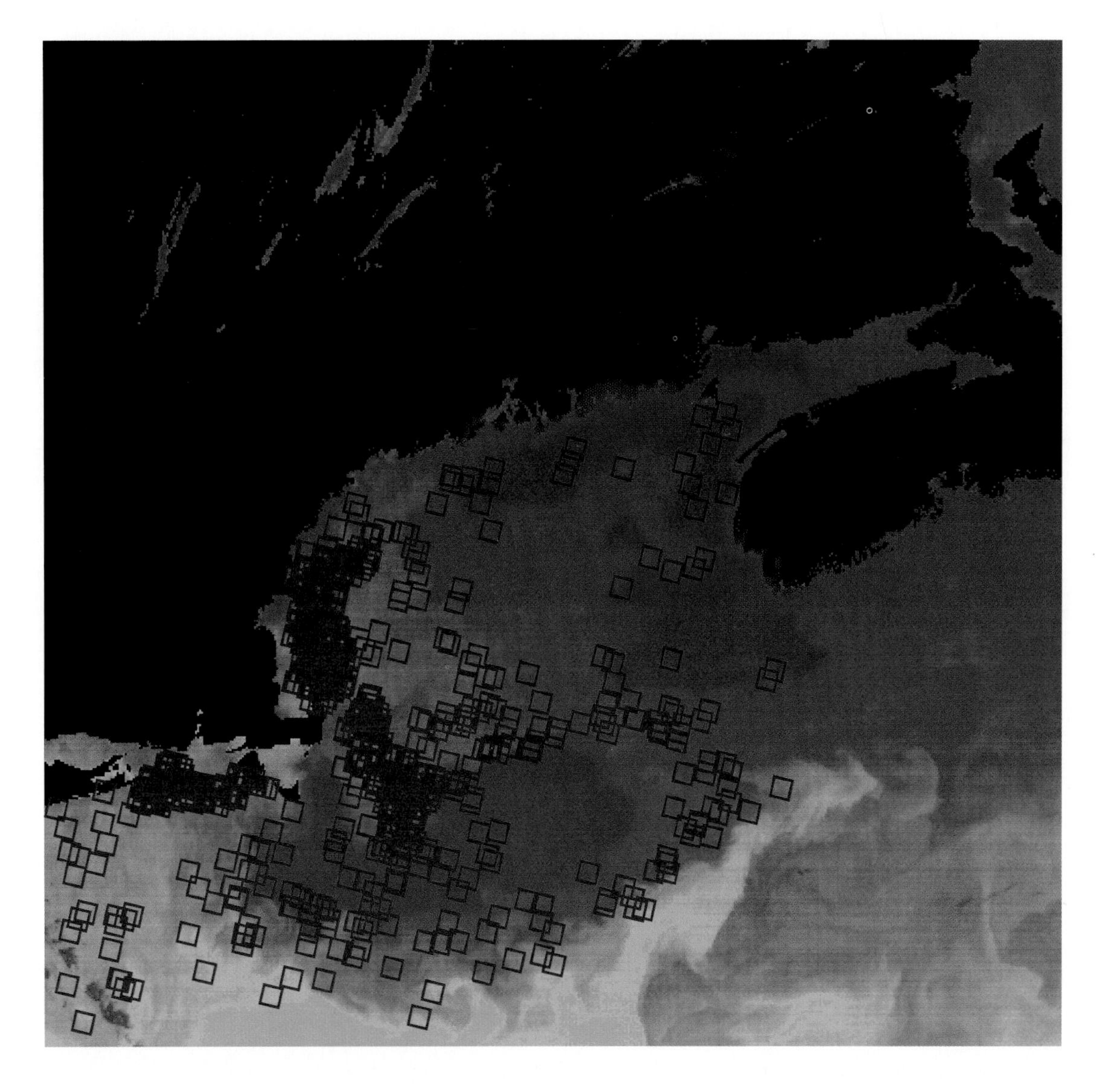

6.3 Finback Whale Distribution

Locations of finback whales have been tracked by researchers at the New England Aquarium in Boston, primarily through the use of aerial surveys. Here all sightings of individual finbacks in the Gulf of Maine and Georges Bank for the 39-month period between November 1978 and January 1982 are plotted on a satellite image of the waters. Finbacks, like other whales, congregate in general locations (such as in Great South Channel near Nantucket) at specific periods of time, although they range widely throughout the entire Gulf. *Credit: data courtesy of S. Kraus, New England Aquarium; image enhancement: S. Meyer, Island Institute*

in chapter 5, whales distribute themselves through different parts of the Gulf according to signals and strategies particular to each species (figs. 6.2, 6.3). Whales therefore act not only to divide up the basic resources of the Gulf, but also to cycle nutrients in a highly efficient manner. They are thus regarded by systems ecologists as an example of a part of the marine system that is spatially integrated; their integrating function lends stability to the system.

The sei whale only passes through the outer rim of the Gulf as a transient bound east or west, while other large whales are seasonal residents commonly seen in coastal waters. The right whale is the most endangered of the Gulf's great whales and the most rarely seen. Most are found off eastern Maine and southern New Brunswick where they feed primarily on swarms of plankton fertilized by local upwellings. Lubec and Grand Manan channels also host other species that are attracted by the schools of herring that gather here to feed and spawn. The shores of West Quoddy and Grand Manan are among the few locations in the world where you might be fortunate enough to see a circle of humpback or fin whales working a cooperative feeding technique to corral a school of herring by means of a bubble curtain.

These whale distributions also suggest their vulnerability to disturbances to particular locations at certain times of the year. An oil spill near the refinery site once proposed for Eastport, for example, could have devastated the few remaining right whales dependent upon the unique properties of this area.

There is a tendency when thinking about the ecosystem of the Gulf of Maine to overlook the presence of marine mammals, perhaps because the whale population today is not large. But this is to ignore the role of mammals in the evolution and functioning of this system when their numbers were much larger than now. We can reasonably guess that the present population of the five large whale species in the Gulf totals about 3,500, while their population prior to colonial exploitation probably totaled about 25,000. Our colleague Spencer Apollonio has suggested that this amounts to a loss of perhaps 75,000 tons of whale biomass from the Gulf. Some of the energy that previously went into the growth of whales undoubtedly appears elsewhere in the form of other marine life. Nevertheless, the absence of so much higher-level animal life must be significant to the ecosystem. Certain functions and constraints once performed by whales have been removed. Unfortunately, there has been little effort to understand the significance of these changes—another failure to think of the Gulf of Maine in ecosystem terms.

Whale numbers continue to fluctuate. Researchers from the New England Aquarium documented an increase in whale sightings in the northern Gulf of Maine in 1993 but they downplayed the significance of the increase, pointing out that sightings on Stellwagen Bank at the Gulf's southern end were lower, and that scientists increasingly view the Gulf of Maine as a destination in whales' far-ranging migrations. Thus more whales may have been attracted to the Gulf of Maine recently because of the availability of food, but the numbers should not be interpreted as a permanent population increase.

New technologies, including improvements in satellite tracking systems, make it likely that researchers will learn more in coming years about the significance of changes in the whale population, and the implications of those changes for the ecosystem as a whole. Satellite tagging, a technology used by terrestrial wildlife biologists for many years, has made it easier to track individual animals (fig. 6.4). Researchers working off Grand Manan recently tracked two right whale journeys throughout the Gulf of Maine; one pair traveled from the Bay of Fundy to New Jersey and back, a journey that took 43 days, while other trips have been as short as two days (fig. 6.5).

Marine Mammals and Man

Even a relatively small population of marine mammals can come into conflict with man. Whales and porpoises become entangled in fishing nets and do thousands of dollars in damage as they

6.4 Right Whale Research
(left)

A medieval invention, the crossbow, enables researchers in the Bay of Fundy to collect skin samples from right whales, an endangered species, to determine their genetic relationships. *Credit: S. Kraus, New England Aquarium*

6.5 Tracking Right Whales
(right)

Based on identification of specific right whales from boats and planes, researchers have followed individuals on their travels throughout the Gulf of Maine. The tracks plotted here illustrate the amount of time different right whales have spent traversing the Gulf. Although straight lines are used to connect specific observations, the routes of these whales through the Gulf are obviously more complex. The shortest trip shown is from near Race Point on Cape Cod to the Bay of Fundy over a two-day period. Other individuals take upwards of five days crossing the Gulf to reach their breeding and feeding grounds off Grand Manan. Trips of longer duration to and from Southeast Channel off Nantucket and back again to the Bay of Fundy sometimes last as long as three months. *Credit: data courtesy of S. Kraus, New England Aquarium; image enhancement: S. Meyer, Island Institute*

attempt to escape; hungry seals—particularly the aggressive gray seals whose numbers have recently increased along the Maine coast—can do substantial damage to nets that enclose fish farms.

Most problems facing marine mammals in the Bay of Fundy arise when they collide—sometimes literally—with humans. Colliding with ships is the single greatest factor in the unnatural deaths of right whales. When right whales are resting at the surface or courting and mating, they are oblivious to vessel traffic. They react only at the last minute, and because they are relatively slow swimmers they are sometimes hit. To make matters worse, they particularly like the shipping lanes to and from the port of Saint John, New Brunswick. Humpbacks are also known to "log" or rest quietly at the surface, creating obstacles for vessels. Finback and minke whales are often found among fishing vessels in areas of the Bay of Fundy. Most of these whales are quick and adept enough to avoid collisions, but collisions do occur. Vessels with engines and electronics shut down are difficult for submerged whales to detect. Fishermen have to be on the lookout when changing directions and moving to new positions in the presence of whales, or they risk running into one.

In 1993, harbor porpoise became the focus of a dispute between the U.S. National Marine Fisheries Service and commercial fishermen. Concerned about *bycatch* of porpoise (a protected marine mammal) in gill nets along the east coast of the United States, the National Marine Fisheries Service called for closures along midcoast and eastern Maine from mid-August through mid-September, a plan designed to reduce porpoise bycatch in this one area by 20 percent, and by 4 percent for the Gulf of Maine as a whole. Under the plan, the length of the closures would double in the second year and triple the third—effectively putting gill net fishermen in the region out of business.

Fishermen responded with new technology: a "pinger" device designed by a marine mammal behavioral scientist from Newfoundland that has been shown in at least one experiment to keep porpoise away from nets. Gill-netters have proposed that the National Marine Fisheries Service allow the substitution of "pingers" for closures to protect harbor porpoise.

Whale watching can also have unintended consequences for marine mammals. Vessels from both Canada and the United States operate in and around Passamaquoddy Bay, as well as on Stellwagen Bank and at other locations in the southern Gulf of Maine. Every year new operators start up. Concern about harassment of whales, even if unintentional, has prompted the Canadian Department of Fisheries and Oceans to develop guidelines for whale watching. Yet operators are still not required to have a license for whale watching, or even to have a knowledgeable naturalist on board.

Whale watching is not simply a commercial undertaking. Powerboats, sailboats and kayaks, and even swimmers and divers approach whales each summer and fall (fig. 6.6). Whales are sometimes harassed, and people put themselves and the whales at risk. Added to these frequent visits are the numerous research projects now under way. Researchers must obtain permits from their respective federal governments if marine mammals are to be captured, marked, tagged, killed, or transported.

Several research organizations are investigating the effects of sound on whales and other marine mammals, particularly the implications of the noisy underwater world man has created (largely through ship traffic) for the animals' ability to communicate with one another. Since 1992 the Bioacoustics Research Program at Cornell University has been working with the U.S. Navy, using its passive acoustic submarine tracking systems to listen to whales in the Atlantic. Researchers have tracked blue whales from distances of over 900 nautical miles (1,650 kilometers) and listened to humpbacks singing from over 50 nautical miles (90 kilometers), concluding that these whales are not only communicating over great distances but using a form of low-frequency sonar in order to navigate, orient, and find areas that might have food.

6.6 Humpback Mother and Calf

This pair of humpbacks was photographed off Seal Island in outer Penobscot Bay, where they were closely approached by a visitor. *Credit: J. Morton*

Harbor and Gray Seals

Harbor seals are the most widely distributed of the 33 species of seals in the world (fig. 6.7). They inhabit the waters and halftide ledges of the shores of Japan and China in the Pacific, the west coast of North America from the Bering Sea to southern California, and both sides of the North Atlantic. The world population is at least on the order of 150,000, and there may be twice as many as that.

Harbor seals are said to be highly intelligent. They are capable of recognizing individual boats from which fishermen have shot at them in the past (most fisherman consider seals to be unmitigated pests); they often "know" whether a fisherman or a salmon farmer has a gun before they can see it, and they can judge rifle range more accurately than most other quarry.

The shores of the Gulf of Maine host something on the order of 30,000 harbor seals, a number that has been steadily increasing on the American side of the Gulf since the enactment of the Marine Mammal Protection Act, one of the nation's early environmental laws that made it a crime to kill, injure, or harass any marine mammal. Although some of the harbor seals of the Gulf of Maine are year-round residents, most swim farther south during the early winter to spend time feeding off the shores of the Atlantic from Cape Cod to the Chesapeake.

They mate in the spring of the year, either in the water or on halftide ledges, and have group ranges that a bull, his harem, and their pups use throughout the summer. The whelping ledges, onto which the females crawl to give birth in May or early June, are generally inshore in protected waters of the region's embayments (fig. 6.8). There is some evidence that seals can control the timing of giving birth, particularly when a storm makes it difficult and dangerous to crawl up on a ledge. If pups are born in the water, their mothers can sometimes cradle them between their flippers, but they must get to land quickly or the pups will drown. Later in the summer they move farther offshore to ledges at the outer rim of their archipelagoes where the feeding is better.

6.7 Harbor Seal Pup on Haulout Ledge
(above)

Each year in the late spring, harbor seals move into protected bays and coves around the rim of the Gulf of Maine to give birth to their pups, which haul out on halftide ledges to rest. *Credit: P. Ralston*

6.8 Harbor Seal Distribution in Midcoast Maine
(right)

Seal haulout ledges (pink dots) in Penobscot Bay are shown in this image based on information collected by the Maine Department of Inland Fish and Wildlife. Researchers believe that as many as 30,000 harbor seals inhabit the Gulf of Maine, where their numbers are still increasing as a result of two decades of protection. *Credit: P. Conkling, T. Ongaro, Island Institute; GAIA image*

The larger, horse-faced gray seals are much less social than harbor seals. They confine themselves to the outermost ledges of a given area and reach their extreme southern limit along the Maine coast. Their numbers increase in the winter when seals from northern Canada migrate into the Gulf of Maine. These are truly enormous seals; the bulls can weigh upwards of 800 pounds. Their total worldwide population probably does not exceed 100,000, which makes them one of the rarer marine mammals on the globe. Like those of harbor seals, gray seal numbers are slowly increasing here, probably as a result of two decades of protection. In March 1994, the first breeding colony of gray seals off the mid-Maine coast was located on an outer ledge that biologists refused to identify to protect them from any would-be marksmen.

Seal populations in the Gulf of Maine and Bay of Fundy are healthy, although worldwide there are serious problems. Particularly in the Bay of Fundy, their abundance—not their scarcity—is causing concern. There is evidence that seals open lobster trap doors or destroy traps to get bait and anything caught inside. They swim freely in and out of herring weirs, tearing holes in twine and driving herring away. They are also accused of stealing bait from fishermen's longlines. Salmon farms serve up an enticing meal, of course. When seals find their way into cages, they leave salmon with gaping wounds, injuring the salmon and potentially introducing infections into a pen with hundreds of thousands of dollars' worth of fish.

For two centuries, the natural function of mammals at the top of the food chain in the Gulf of Maine has been vulnerable to the disturbance of their natural regulatory function within the system—first by overzealous hunting, then by overzealous protection. Whales and seals were pursued virtually to extinction for their commercial value, and because it was believed (at least in the case of seals) that they competed with the fishing industry. When protection came in the form of the Marine Mammal Protection Act, it was absolute: in U.S. waters, killing a whale or seal, or even taking it into captivity without a permit, is strictly prohibited. Seals have now proliferated to the point where, in the Bay of Fundy and parts of eastern Maine, some fishermen regard them as a serious nuisance.

Shorebird Migrations

For decades ornithologists have known that the immense tidal flats surrounding the Bay of Fundy are important feeding stops for migrating shorebirds (fig. 6.9). Semipalmated sandpipers, for instance, which breed in summer in northern Canada and winter in South America, stop for a couple of weeks on these flats and gorge themselves primarily on worms and intertidal crustaceans, storing up fat reserves to sustain them for the long flight south. They increase their body weight by nearly one-third during this period of intensive feeding. Between one and two million shorebirds stop over annually in Fundy, including 50 percent of the estimated world population of semipalmated sandpipers. In Shepody Bay at the entrance of the Bay of Fundy, nearly half a million shorebirds may be present at one time during midsummer (fig. 6.10).

In 1994, when shorebird numbers dropped in Shepody Bay, some speculated that siltation caused by the causeway on the Petitcodiac River above Shepody Bay had made its way to the area and had destabilized the upper layer of mud in which the birds' food supply (*Corophium*) is found. It seemed to be a local phenomenon since there was no decrease in shorebird visitations in Nova Scotia just across the bay. Something had happened on the New Brunswick side to affect the *Corophium*: they just weren't there.

A widespread problem for sea and shore birds is the physical destruction or alteration of their habitat. This happens gradually, a little bit here, a little bit there, but the cumulative small losses have had major effects. Hundreds of thousands of northern and red phalaropes have virtually disappeared from the passage at the mouth of Passamaquoddy Bay in recent years. This phenomenon is currently being studied and might be

6.9 Shorebird Staging Areas in the Upper Bay of Fundy

This image shows thousands of square miles of mud flats (yellow) where sandpipers, plovers, and other shorebirds congregate in immense flocks to gorge on tiny crustaceans in the flats before making the long flight across the Gulf of Maine to Cape Cod and points south. Upwards of 500,000 individuals will gather here on their southern migration during late July and early August. *Credit: T. Ongaro, P. Conkling, Island Institute; GAIA image*

6.10 Shorebird Flock

Part of an immense flock of semipalmated sandpipers at a roost site, Mary's Point, New Brunswick, one of two Western Hemisphere Shorebird Reserves in Canada, considered critical to the shorebirds' migratory success. *Credit: S. Homer*

linked to pollution problems. Islands that for thousands of years had been used by seabirds as nesting sites have been occupied by people, and the birds either hunted or driven out. This has resulted in great declines in some cases; gannets and great cormorants have stopped nesting in the Bay of Fundy completely. Shores and beaches where birds nest, feed, or roost have been taken over for wharves, fish plants, dwellings, or recreational areas. Beach-goers disrupt nesting, roosting, or feeding birds. Motorized vehicles destroy beaches, sand dunes, and marshes. Tidal marshes have been altered and destroyed by drainage for agricultural land or more recently by flooding for waterfowl management. There have been proposals to generate large amounts of electrical energy by constructing tidal dams across part of the Bay of Fundy. If the mud flats above these dams were permanently flooded, thousands of shorebirds would be deprived of a key source of food on their southward migration.

Salt marshes and seagrass beds are important in the ecology of shorebirds and several varieties of waterfowl. Canada geese and such shorebirds as greater and lesser yellowlegs and least sandpipers forage in the extensive salt marshes along the fringe of Cape Cod and the islands of southern New England. Farther north, they prospect for food in smaller marshes during their northward migrations in spring.

Seabirds: The Perils of Specialization

Seabirds abound in the Gulf of Maine, but they are found primarily on the smallest and most offshore islands and ledges. Of the more than 5,000 islands in the Gulf of Maine, only about 10 percent offer the right combination of conditions to be suitable for gulls, terns, cormorants, petrels, guillemots, and the other species that comprise the region's diverse seabird community. What is special about a seabird island that makes it so rare? First, it must be entirely free of large mammals, because seabirds cannot protect themselves from any mammal capable of eating eggs, young birds, or adults. Satisfying this criterion alone is no simple task in the Gulf of Maine, because foxes, mink, coyotes, raccoons, and other predatory mammals make regular appearances on most of the Gulf's islands (winter ice offers them access to these islands, and they are frequently able to persist year-round on the larger islands). Second, most species of seabirds prefer an open habitat free of forest growth for nesting. Last, a seabird island must offer quick access to offshore food sources.

The prime places in the Gulf of Maine that meet these three conditions are the outermost islands and rocks that ring the Gulf. Islands such as Machias Seal off Grand Manan, the Libby and Brothers islands off Machias, Schoodic and Great Duck off Mount Desert, Matinicus Seal, Matinicus Rock, and Metinic Islands in outer Penobscot Bay, and sandy islands off Cape Cod such as Manomet and Monomoy collectively support tens of thousands of nesting pairs of seabirds (fig. 6.11).

In addition to being geographically specialized, Gulf of Maine seabirds have experienced several periods of dramatic decrease due to human hunting. The early records of voyageurs, fishermen, and naturalists give a hint of the original mix of species and the size of their populations. From these records we know, for instance, that herring and black-backed gulls, now among the most abundant species, were historically very rare, while terns and alcids (puffins, guillemots, murres, and auks) were more abundant. We also know that many seabird species in the Gulf of Maine and throughout the North Atlantic were hunted nearly to extinction in the late 1800s, and that the great auk and Labrador duck did not survive at all.

Aggressive hunting for east coast markets reduced eider duck numbers to near extinction along the Maine coast. According to ornithologist A. H. Norton, it was not unusual for a pair of hunters to shoot several hundred eider in a single morning. By 1904 there were only four records of eiders nesting on the Maine coast, near Old Man Island in remote Machias Bay. Gannets and cormorants were shot and the young chicks used for cod bait. Few local fishermen mourned the near extinction of

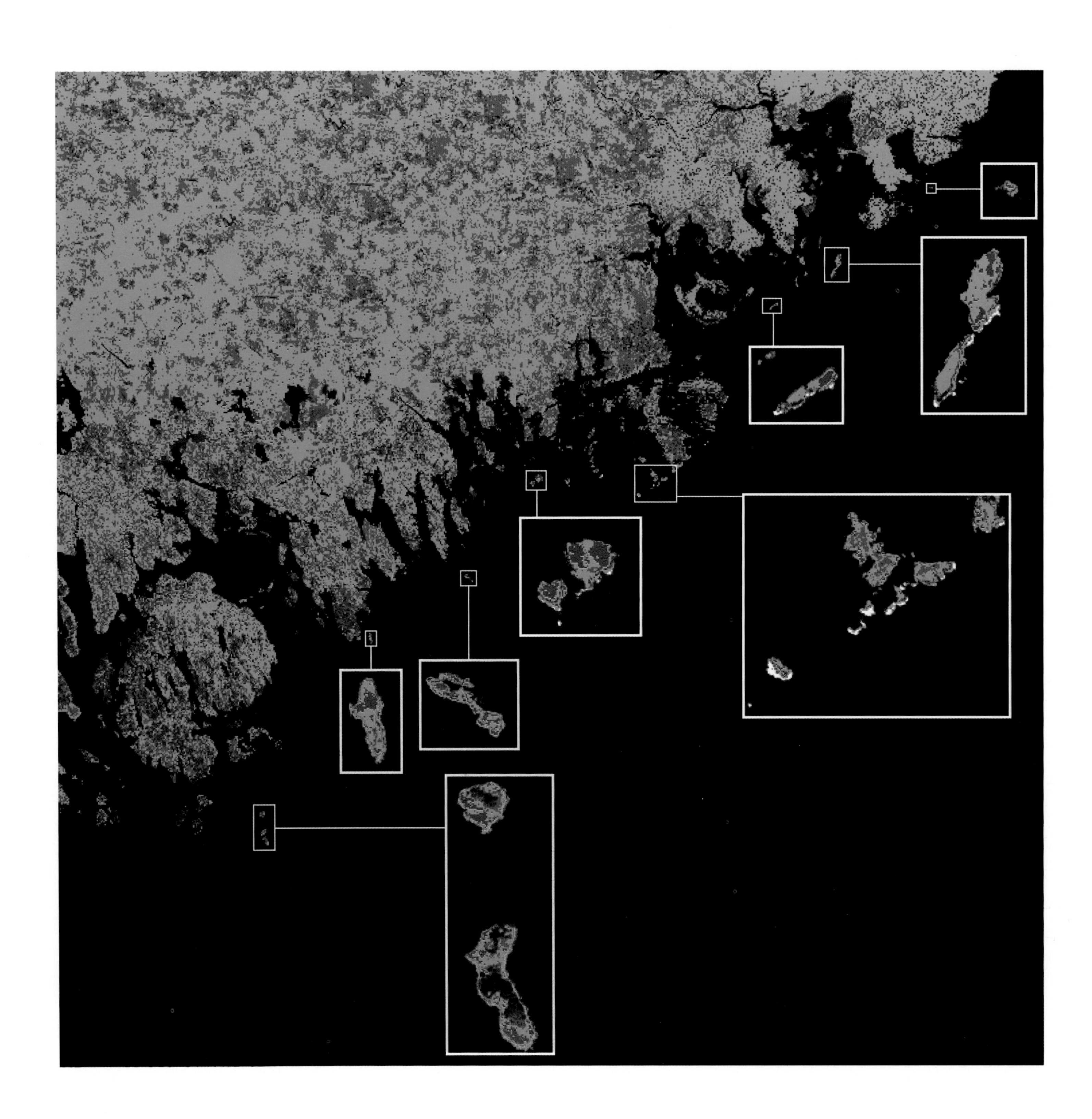

6.11 Seabird Nesting Islands along the Eastern Maine Coast

Seabirds such as gulls, cormorants, guillemots, terns, and eiders key in on isolated islands free of predators and near productive feeding grounds. Seabirds prefer a mix of open ground and shrubby vegetation for nesting sites, which researchers can locate and quantify as potential habitat through the use of satellite imagery. Large nesting colonies are found in a band along the eastern Maine coast near areas of vertical mixing (see figs. 5.4 and 5.7). *Credit: R. Podolsky, P. Conkling, Island Institute; GAIA image*

cormorants from most areas in the Gulf of Maine by 1900, believing that cormorants competed with them for the same fish.

In addition to gunning and egging for food and bait, there was a very active Gulf of Maine market to supply feathers to the millinery trade. Everything from seabird heads to the wings and tails of terns, gulls, and puffins was being used to adorn hats as far away as San Francisco and Asia. It was during this period that the Atlantic puffin was eradicated from all but two Gulf of Maine islands—Matinicus Rock and Machias Seal.

Protection undertaken in the early 1900s has allowed seabird numbers to rebound in the Gulf of Maine and throughout the North Atlantic. By 1931 eider ducks began to reappear around many of their long-abandoned nesting islands in the Gulf. Now they are almost superabundant, and rafts of tens of thousands are not uncommon. Cormorants have recovered even more dramatically, and by 1940 many former colonies contained thousands of individuals.

Nonetheless, sanctuaries and treaties are unfortunately not enough to guarantee the viability of any population. Pollution, plastic debris at sea, intense fishing pressure, and coastal landfills have all undermined the progress made during the middle of this century. For example, by the mid-1960s it became apparent that herring and black-backed gulls, which have learned how to scavenge along sandy dunes of Cape Cod and from fishing boats, were experiencing a dramatic recovery that was displacing the less aggressive Arctic and common terns (fig. 6.12). In 1901 there were only 14 herring gull colonies in the Gulf of Maine; now there are over 270 colonies between Cape Cod and Maine alone, supporting over 35,000 nesting pairs.

Recent successful efforts to restore populations of Arctic terns, Atlantic puffins, thick-billed murres, and Leach's storm petrels have helped maintain healthy populations of these most vulnerable inhabitants of the Gulf.

6.12 Herring Gulls and Fishing Boats

One of the reasons that herring and black-backed gulls have been so successful is that they readily adapt to new food sources and learn to feed off fishing wastes and the flotsam left behind on beaches. *Credit: P. Ralston*

Terns: The Case for International Cooperation

Terns are small, nimble-flying members of the seabird order that includes gulls, shorebirds, and alcids (puffins and their relatives). They are light-colored and long-winged, with deeply forked tails and black caps (fig. 6.13). Terns arrive in the Gulf of Maine in mid-May and depart for their wintering grounds at sea in August and September. By early June most have mated and are busy incubating from one to three eggs. By late June these eggs begin to hatch. Three weeks later the chicks are learning to fly. In mid-August the parents and fledglings leave the colony and begin their long migration southward. The Arctic tern holds the distance recordfor any migrating bird, traveling over 20,000 miles from the Gulf of Maine south to Antarctica and back again each year.

Terns are probably the most fragile community of birds inhabiting the Gulf of Maine. An international group, the Gulf of Maine Tern Working Group, was founded in 1987 with the purpose of halting the downward trend in their populations. A model for cross-border cooperation, the Tern Working Group's 50 members include Canadian and U.S. biologists and birdwatchers with provincial, state, federal, and nonprofit affiliations. The group coordinates an annual survey of every island in the Gulf that has ever supported terns.

Meeting late each summer, this informal Canadian-American group reduces the region's latest tern population surveys to numbers on a blackboard. Special attention is paid to common, Arctic, and roseate terns, but the group also keeps an eye on the population fluctuations of least, black, and Forster's terns, as well as piping plovers and skimmers.

Perhaps more than any other organism, terns are predictors of the health of the marine ecosystem. They eat small fish that they glean from the top meter of ocean by plunge-diving from the air. They feed their young mostly two-to-three-inch herring or hake, with smaller amounts of sand eels, butterfish, and baby bluefish. When food is meager around a nesting island,

6.13 Arctic Tern

Arctic terns nest in dense colonies on a small number of islands in the Gulf of Maine. This tern is bringing a small herring to its young on Matinicus Rock in outermost Penobscot Bay. *Credit: P. Ralston*

terns are forced to feed their young lower-quality crustaceans, which can lead to starvation and low reproductive output.

Geographically, terns concentrate at fewer than 30 islands in the Gulf of Maine. The roughly 8,500 pairs of terns counted there in 1993 took up less than 10,000 square meters (two football fields) in habitat, and geographic isolation is their major habitat requirement. Development and disturbance must be prevented on the few remaining acres of sandy beaches where least terns nest if they are to survive (fig. 6.14).

Over the 90 years that terns have been monitored, they have ridden a roller coaster of precipitous population crashes and recoveries. Since the Gulf of Maine Tern Working Group's inception, most terns have experienced population upswings. Common terns numbered 4,300 pairs in 1993, up from 2,100 pairs in 1987. Arctic terns numbered 4,400 pairs in 1993, up from 3,100 pairs in 1987. Roseate terns numbered 142 pairs in 1993, up from 52 pairs in 1987. All told, terns are up about 60 percent over their 1987 numbers.

Human intervention by hunters, polluters, and protectors has had profound effects on wildlife populations in the Gulf of Maine. So have human-induced changes in the food supply and the destruction of habitat through development. Protecting one species or group of species—for example, removing gulls to encourage terns or prohibiting the killing of seals—invariably affects other species in the ecosystem. The filling of wetlands, the development of beaches, and the introduction of pollutants in the marine environment can all be counted on to reverberate through the system from top to bottom.

Two centuries ago, human beings replaced whales as the top predators in this marine system; today, even as we enact laws to protect whales, seabirds, and other endangered species, decisions made by humans will determine the character of the entire ecosystem for centuries to come.

6.14 Least Tern Habitat

Least terns are vulnerable in the Gulf of Maine because their preferred nesting habitat is on sandy beaches (top)—the same place where millions of tourists and vacationers head. In addition, these beaches are subject to significant natural disturbances such as the winter storms (bottom) that can not only alter their habitat but carry it away offshore. *Credit: image enhancement: R. Podolsky, Island Institute; data courtesy of K. Fink*

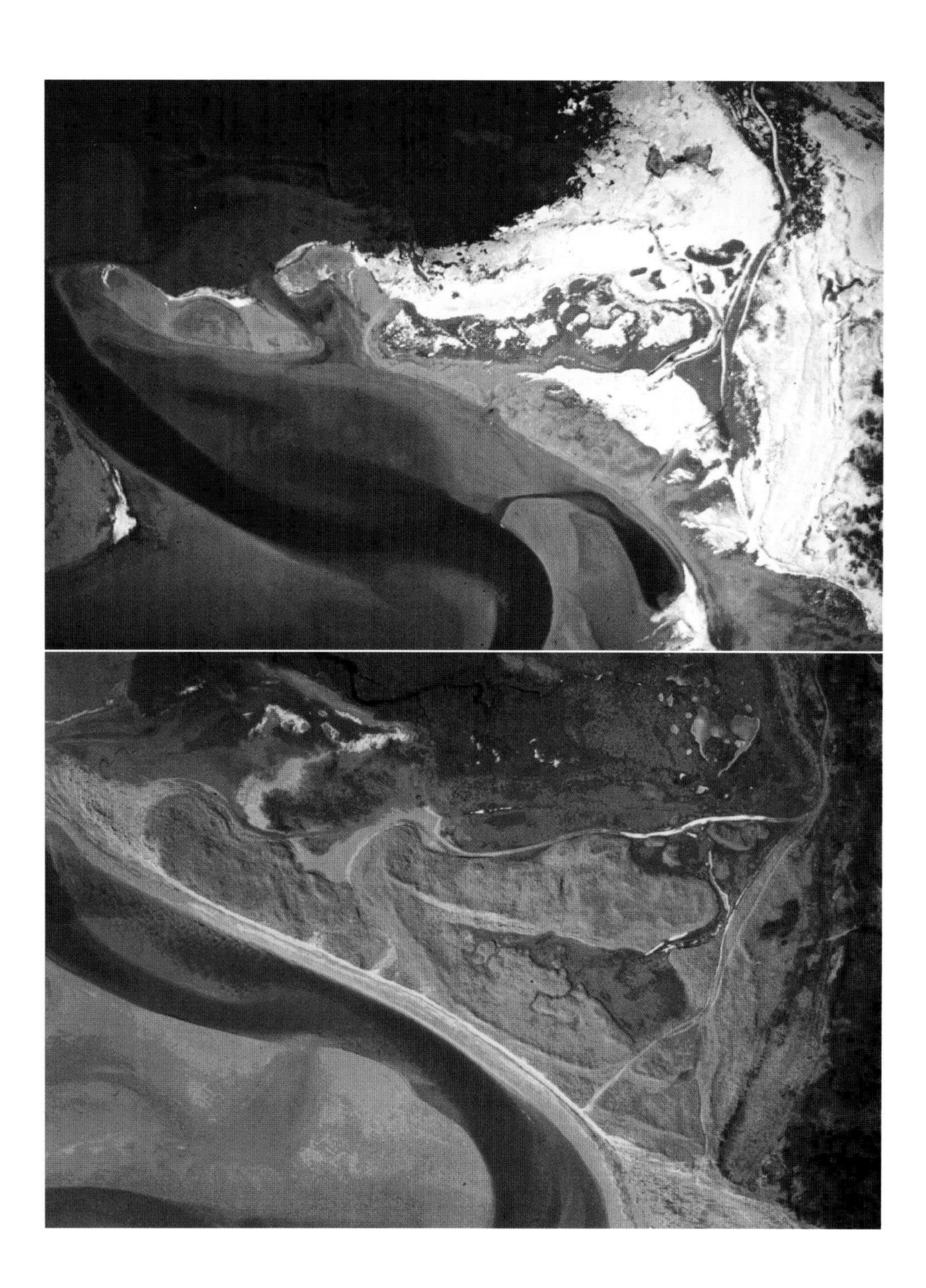

PART TWO: MOVING ASHORE

7.1 Inshore Marine Habitats

The rocky intertidal community of the Gulf of Maine is spatially complex and biologically diverse. *Credit: P. Ralston*

7

THE THIN EDGE BETWEEN LAND AND SEA

Janice Harvey, Kenneth Mann, Richard Podolsky, and Philip W. Conkling

Of all the places on earth, shorelines are among the most ecologically dynamic and productive. Here at this "thin edge," as one writer termed it, fresh and salt water, wave, wind, and land commingle with an array of specially adapted plant and animal species to create zones of immense biological productivity. Salt marshes, submerged seaweed and seagrass beds, and rocky and barrier islands all figure prominently in enhancing the productivity of nearshore and estuarine waters. As such they must be viewed as a distinct and essential part of the ecology of the Gulf of Maine region (fig. 7.1).

An important distinction between the biology of offshore and inshore waters concerns the plant life at the base of the food chain. Offshore, life depends on microscopic single-celled plants, or phytoplankton. Closer to shore, both between tide marks and for some distance below low tide, larger, highly visible salt marsh plants, seagrasses, and seaweeds grow attached to the bottom, forming dense beds that are vitally important nursery grounds for a very large percentage of the Gulf's fish and shellfish species.

The Critical Edge Where Fresh and Salt Water Mix

Estuaries are distinctive areas found wherever a significant supply of fresh water meets the sea. The immense terrestrial watershed of the Gulf of Maine empties 250 billion gallons of fresh water into nearshore estuarine environments each year (figs. 7.2, 7.3). This fresh water carries suspended and dissolved nutrients that enhance the productivity of the estuarine environments. Here we find a variety of coastal wetlands such as salt marshes, mudflats, and gravel and rocky areas that utilize the nutrient flows from the river discharges. Like other wetlands, they filter and clean water before it enters the marine portion of the Gulf of Maine system. Estuarine waters are not as salty as the open bay or as fresh as the river upstream, and thus a different mix of plants and animals live there.

Although estuaries are comparatively rare worldwide, making up less than one percent of the globe's shorelines, the Gulf has a relative abundance of them, which in part accounts

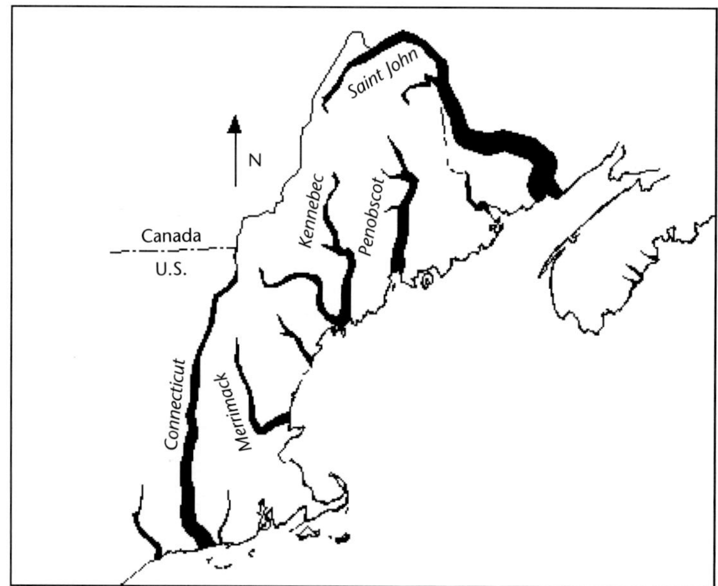

7.2 Freshwater Rivers of the Gulf of Maine
(above)

The major and minor rivers of the Gulf of Maine are shown (in pink); the Saint John is the longest, followed by the Penobscot and then the Kennebec. *Credit: S. Meyer*

7.3 Freshwater Inflow
(right)

Freshwater runoff into the Gulf of Maine is shown from the region's major rivers. The illustration is a graphic representation of the relative scale of each river's average annual runoff contribution to the Gulf. All the rivers together deliver an average of approximately 250 billion gallons of fresh water each year, with the Saint John contributing nearly a third of the total.

The estuaries of these rivers are nurseries and feeding grounds for many of the fish and shellfish of the Gulf. The special mixing of waters that results when fresh water flows into seawater concentrates nutrients and encourages the production of food suitable for the immense numbers of larval and juvenile sea life that are sheltered in these estuaries. *Credit: data: S. Apollonio; illustration: T. Christensen*

for the richness of the entire Gulf. According to the estimate of the Gulf of Maine Council on the Marine Environment, nearly one-third (22,417 square miles) of the international watershed is estuarine.

Among the major estuaries in the southwestern part of the Gulf are Massachusetts Bay and Great Bay in New Hampshire (fig. 7.4). The long Maine coast supports the largest number of estuaries; west to east, important ones are found in Saco Bay, Casco Bay, Merrymeeting Bay, Sheepscot Bay, Muscongus Bay, Penobscot Bay, Blue Hill Bay, Frenchman Bay, Narraguagus Bay, Englishman Bay, Machias Bay, Cobscook Bay, and Passamaquoddy Bay which straddles the international border (fig. 7.6). In Canada one of the most important estuarine areas of the Gulf is associated with the region around the mouth of the Saint John River, which drains the largest land area and carries the largest volume of fresh water to the sea. In the upper Bay of Fundy, around the edges of Shepody and Chignecto bays where meandering rivers meet the sea, are the most extensive estuarine areas of the Gulf. In some senses, you could say that the entire Bay of Fundy is an estuary and not be far from the mark.

Pieces of the Estuarine Environment: Salt Marshes

A salt marsh is a fringing coastal wetland, a thin ribbon of hay and grasses that is connected to the marine environment briefly at high tide. During periods of flooding, marshes that accumulate nutrients during the growing season release them into the sea and thereby enrich nearshore waters. Salt marshes are particularly common in estuaries, where silt brought down the rivers and deposited in these quiet waters provides ideal soil for marsh plants to root. A salt marsh is inhabited by a variety of flowering plants that can tolerate having their roots in salt water. The most common is marsh grass, or *Spartina*. Once a bed of marsh grass starts to grow, it acts as a natural trap for silt and makes it possible for the area of the salt marsh to increase steadily.

7.4 Piscataqua River Marsh and Estuary
(following page)

Tidal wetlands of estuaries and embayments are some of the most ecologically important and sensitive nearshore environments within the Gulf of Maine. They serve as habitats for numerous marine and terrestrial organisms, provide nutrients to marine ecosystems, act as filters of pollutants and buffers against storm surges, and are important aesthetically. Because of increased pressures on tidal wetlands within the Gulf due to a growing coastal population and associated development, it is important to carefully evaluate, map, and characterize them.

From 1990 to 1992, detailed mapping and inventories of tidal wetlands were conducted within the Great Bay/Piscataqua River estuary, located on the boundary between New Hampshire and Maine. A combination of remote sensing and extensive field verification was used to delineate the upland and seaward boundaries of these wetlands, plus the boundaries between high and low marshes and seaweed (fucoid) vegetational limits. The viability and status of these intertidal vegetated habitats was also assessed.

Color infrared (CIR) aerial photography supplied the remote sensing component of the mapping. CIR film causes the image to be false-colored. For instance, in this photograph the forest vegetation shows up as various shades of red, the marshes are a brownish to pink color, the seaweeds are a bright red, and the water is black. Because of the color differences, marshes can be discerned and measured apart from terrestrial upland vegetation.

To maximize the quality and amount of information obtained, the overflights were conducted during periods of spring low tides and very clear atmospheric conditions. Additional overflights were done after the first fall fronts in order to maximize color differences in vegetation on the CIR photographs. Vegetational components have a more homogeneous color during the peak summer growing period, while the differential browning and drying of plants in fall enhances the color separation among plant types (and thus the identification of vegetation boundaries). For example, saline types of vegetation tend to deteriorate more slowly than freshwater plants.

The CIR photographs provide valuable information concerning the present conditions of these wetlands and serve as a baseline against which future changes can be assessed. The mapping project has established reasonable assessments of acreage of marshes and seaweeds, as well as relative determinations of high versus low marshes. In addition, the photography and mapping products are being used to determine historical changes (erosion or accretion) of tidal wetlands within the Great Bay/Piscataqua River estuary. *Credit: L. Ward, Jackson Estuarine Lab, New Hampshire*

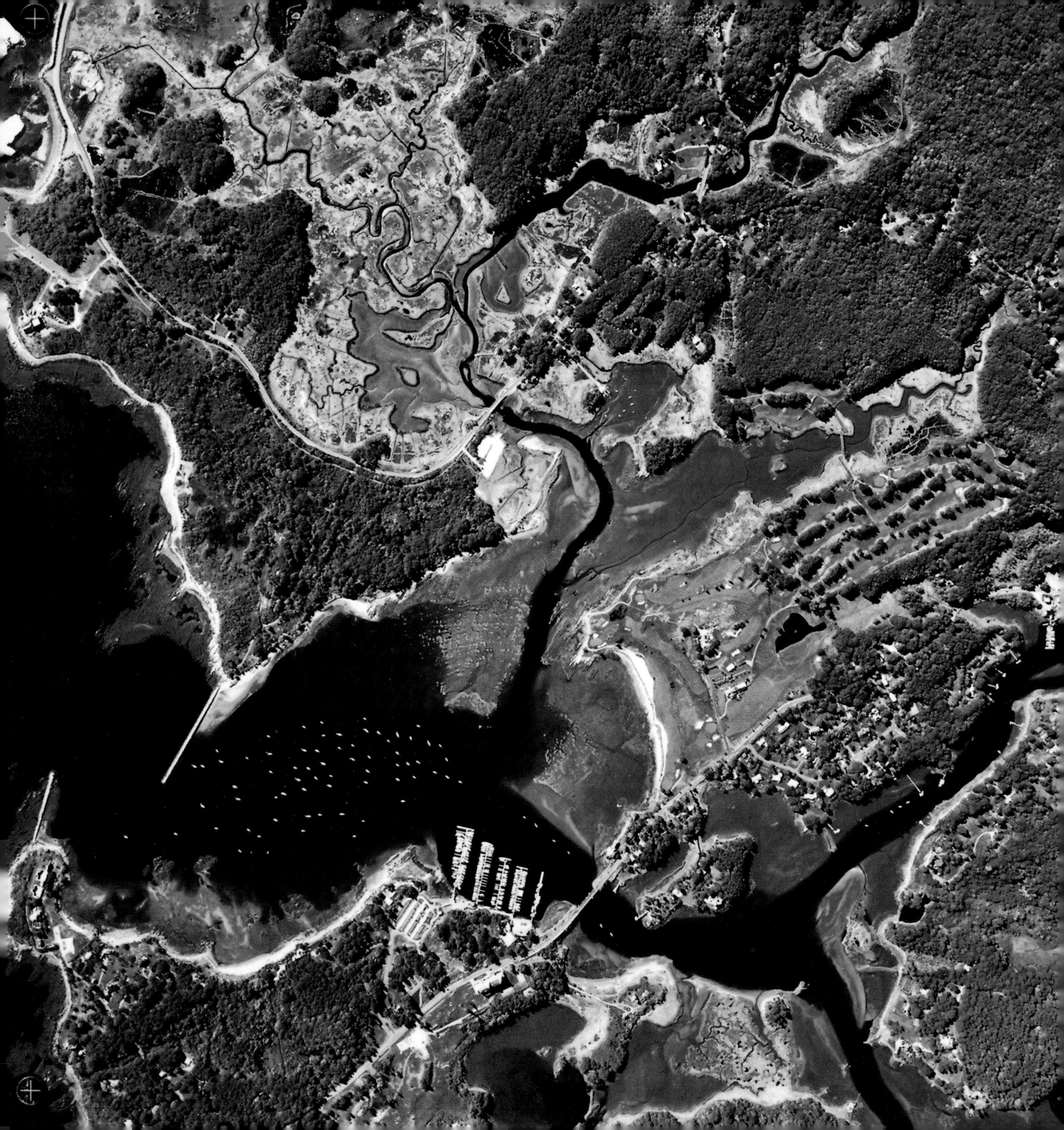

7.5 Salt Marsh Aerial

This view of a salt marsh in the southern Gulf of Maine shows the characteristic branching and twisting of watercourses that flow in and out of the marsh on the tides. *Credit: P. Conkling*

7.6 Sheepscot River Estuary

This subtidal view shows the diversity of the bottom habitats where the Sheepscot River empties into Sheepscot Bay, creating an estuary long known as an important inshore cod spawning ground and productive lobster-fishing area. Yellow and blue areas are subtidal sand and gravel habitats, orange areas are rocky habitats, and purple areas denote muddy bottom. Besides the variety of substrate types, the topography of the bottom adds further complexity to these habitats. *Credit: R. Langton, Maine Department of Marine Resources*

Aerial views of any of these marshes show characteristic branching channels, or creeks (fig. 7.5). The rising tide fills these creeks and the ebbing tide empties them again. It is as if the salt marsh is breathing in and out. Each fresh flow of seawater brings supplies of fertilizing substances, such as nitrates and phosphates, to stimulate plant growth. Other things being equal, the annual growth of marsh grass is proportional to the amount of seawater entering and leaving with each tidal cycle. It is greatest at places where there is a large rise and fall of the tide, and less where the tidal range is small. Restricting the tidal flow by the construction of a causeway or dam has the effect of reducing the productivity of salt marshes.

There are particularly large salt marshes occupying hundreds of acres at the head of the Bay of Fundy and around the perimeter of Cape Cod. Extensive salt marshes are also associated with the mouths of the major river systems entering the Gulf of Maine such as the Penobscot, Kennebec, and Piscataqua. Some salt marshes, like those found around Passamaquoddy Bay, are small pockets of a few acres. Others, like Saints Rest Marsh in the city of Saint John or Scarborough Marsh south of Portland, Maine, are hundreds of acres in size (fig. 7.7).

Salt marshes provide critical nursery grounds for virtually all of the commercially important species of fish and shellfish harvested in the Gulf. To take one example from the upper Bay of Fundy, the summer feeding grounds for the entire Atlantic shad population on the eastern seaboard of the United States and Canada appear to be supported by the production of the bay's salt marsh food web. Large populations of juveniles of other important commercial groundfish such as cod, hake, and pollack spend their winters at the edge of salt marshes, where food like shrimp is found in abundance.

Though the salt marshes have been built over thousands of years, creating a major source of organic carbon for the marine system, it has taken just 300 years of human intervention to drastically change the landscape. Today, for example, only 15 percent of the original Bay of Fundy marshes remain. Fundy salt marshes have been turned into fertile agricultural land using a system of dikes and gates to halt the flow of tidal water onto the land. The single largest impact was made by early Acadian settlers in the upper bay. "Tintamarre" was the name seventeenth-century Acadian settlers gave to what we now call the Tantramar Marsh region at the head of the Bay of Fundy. It means great or thundering noise, describing the sound of the millions of migrating waterfowl that visited the marshes every year (fig. 7.8). With diking, marsh plant and animal life was altered and the flow of nutrients from the marshes to the bay ceased. Much of this land remains blocked off from the sea. However, in the past 25 years some of the converted agricultural lands have been abandoned and freshwater wetlands have been created in their stead. Whether as freshwater wetlands or as salt marshes, it is imperative that these significant resources of this region are understood, valued, and preserved.

Seagrass Beds

Seagrasses are also flowering plants that have evolved the ability to live in salt water. Like *Spartina*, seagrasses need silt or silty sand for their roots. In contrast to the salt marsh plants, they can remain totally submerged for long periods, even indefinitely. The most common seagrass in the Gulf of Maine is eelgrass, *Zostera*. It occurs in the zone above low tide and extends down to several meters below low water. Eelgrass beds often occur near salt marshes, starting at the edge of the salt marsh and extending deeper until it is too dark for them to grow (fig. 7.9). The presence of the seagrass slows the currents passing through it, so that more silt is deposited and the bed tends to grow. Both salt marshes and seagrass beds are important in protecting the coastline from erosion by storms, waves, and currents. Earlier generations used to collect dried eelgrass from the beaches and use it for insulation around their houses. In the 1950s it was even manufactured into panels that could be built into the walls of new homes.

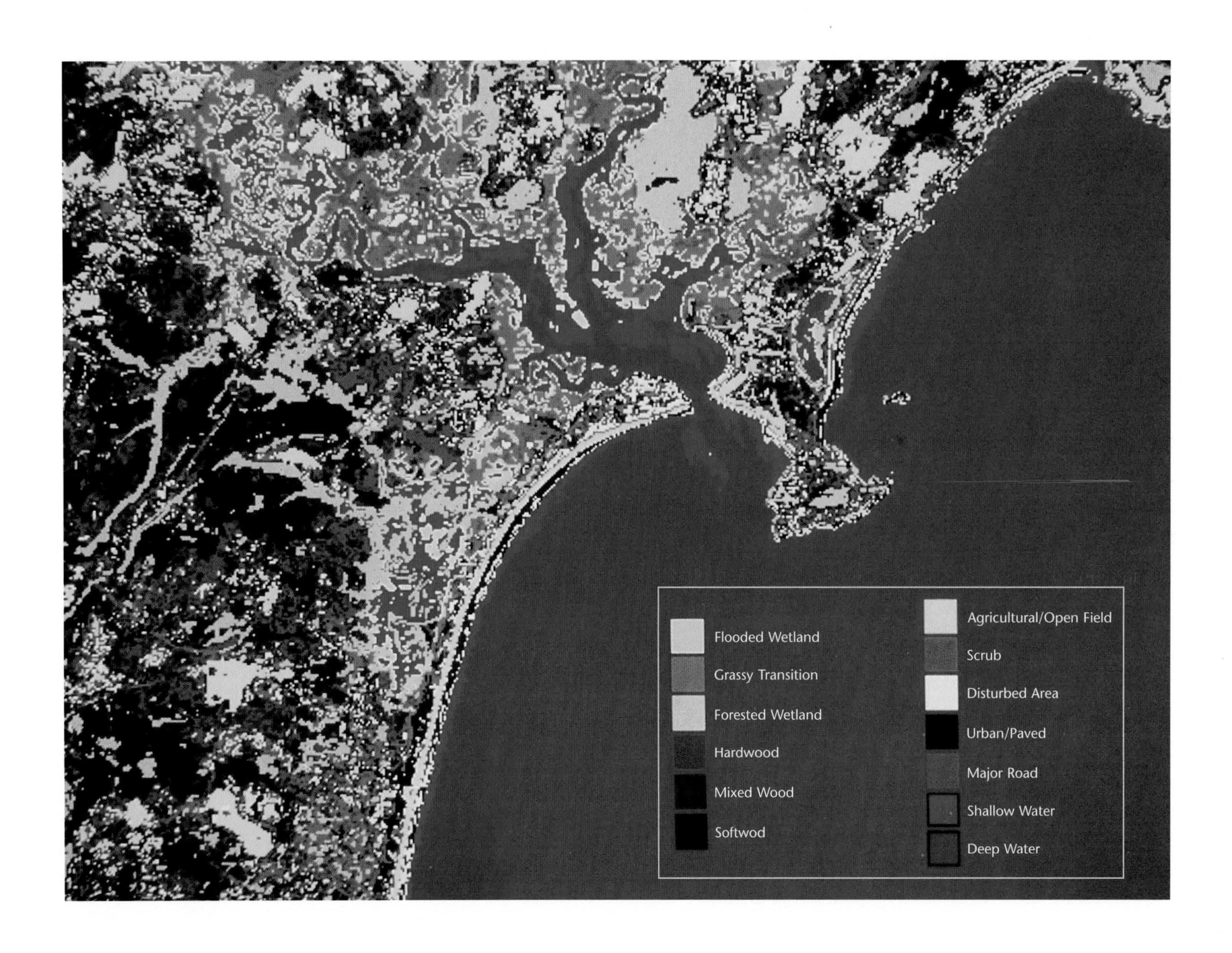
Flooded Wetland
Grassy Transition
Forested Wetland
Hardwood
Mixed Wood
Softwod
Agricultural/Open Field
Scrub
Disturbed Area
Urban/Paved
Major Road
Shallow Water
Deep Water

7.7 Scarborough Marsh Habitats
(left)

This satellite image theme map of the region around Maine's largest salt marsh at Scarborough shows the complex patterns of marine and freshwater wetlands as they intergrade into one another. *Credit: R. Podolsky, P. Conkling, Island Institute; GAIA image*

7.8 Tantramar Marsh
(right)

This marsh, originally settled by French Acadian farmers, is the largest salt marsh in the Gulf of Maine. Its annual production, measured as the biomass of marsh grass growth each year, feeds complex intertidal and subtidal food webs and is a significant contributor to the nutrient charge of the upper Bay of Fundy. Marsh grasses are shown in yellow and comprise approximately 45 square kilometers (17 square miles or over 10,000 acres). *Credit: T. Ongaro, P. Conkling, Island Institute; GAIA image*

Diked Marshland

Shallow Water / Intertidal Zone

7.9 Nantucket Eelgrass Beds

True-color negative film is highly effective in discerning submerged seagrass areas. In this case, the dense seagrass community (dark green) is clearly seen near the jetty at the entrance to Nantucket Harbor at 20 feet (6 meters) below mean low water.

Submerged seagrass vegetation plays an important role in sheltering and supporting many marine finfish and shellfish species in coastal environments throughout the Gulf of Maine. Seagrasses are vulnerable to destruction due to coastal dredging and filling projects, and are also impacted by degraded water quality due to land use runoff. Much of this submerged seagrass has already been destroyed; continued loss will compromise the productivity of coastal ecosystems.

An accurate mapping inventory of seagrass vegetation areas is critical to preserving the high productivity of this important resource. Most features that appear on the photo need to be verified in the field to insure proper classification. In interpreting these photographs, for example, seagrass must be distinguished from marine algae. Accurate seagrass mapping can provide state, provincial, and federal regulatory agencies with information to devise effective protection programs, and will serve as a baseline from which to measure future changes.

The seagrass data shown in this photograph are part of a statewide wetlands mapping inventory being conducted by the Massachusetts Department of Environmental Protection's Wetlands Conservancy Program. *Credit: C. Costello, Section Chief, Massachusetts Department of Environmental Protection*

Along with salt marshes, seagrass beds serve as vitally important nursery areas for a large number of species of fish and shellfish that are harvested from the region.

Seaweed Beds

Seaweeds are found where the water movement is so strong that sediments are mostly carried away, leaving a habitat substrate of bedrock or boulders. Seaweeds are not flowering plants like eelgrass, but rather are algae. They do not have roots but absorb their nourishment from the water washing over the surface of their fronds. Most seaweeds are fixed firmly to the bottom, held down to the rock by special structures call *holdfasts* (fig. 7.10). Most of them cannot flourish where there is soft sediment.

The most common seaweeds in the Gulf above the low-water mark are rockweed (*Ascophyllum*) and bladderwrack (*Fucus*). The latter is easily recognized by the air-filled bladders strung along the length of the blade, which serve to keep it near the surface as the water rises over it. *Fucus* tends to dominate in areas more exposed to wind and waves, while *Ascophyllum* thrives in more sheltered places. In places where rockweed is plentiful, it may be harvested and used to extract alginate, a form of gum that is useful for stabilizing ice cream and cosmetics.

The various ecological functions of the extensive rockweed beds that blanket the rocky intertidal zones along the Gulf of Maine are just beginning to be understood. At high tide their extensive fronds, up to two decades old, float at the water surface over thousands of acres to create a miniature forestlike canopy under which young fish such as tomcod, pollack, sculpin, gaspereau, white hake, cod, and winter flounder find shelter from light and predators. At low tide the fronds collapse heavily on the rocks, providing a wet mat of protection for countless tiny creatures left high and dry by the receding tide. Over 15 species of birds and at least 22 species of fish utilize rockweed forests during some part of their life cycle. Rockweed also undergoes a process of decomposition that is similar to the decay of

7.10 Rockweed Fronds

Rockweeds are various species of marine algae that densely cover rocky intertidal and subtidal zones of the Gulf of Maine, providing cover for numerous invertebrates, oxygenating the water, and serving as food for numerous smaller species when they break loose and decompose. *Credit: P. Ralston*

leaves on a forest floor. Its fronds become part of the detrital food chain that produces food and nutrients to the marine ecosystem. Beginning in May and peaking in August and September, portions of the rockweed plant begin to break off and drift out to sea, forming extensive mats up to a half-mile (one kilometer) in circumference. Beneath and within these mats, large numbers of tiny animals (zooplankton) aggregate. These animals in turn become food for larval lobsters and larval and juvenile fish such as lumpfish, sticklebacks, four-bearded rockling, and red hake. The mats also attract seabirds such as phalaropes and terns that feed on the fish and zooplankton.

The most spectacular growths of seaweed are visible only to divers. On rocky shores, from low-tide level down to depths of about 60 feet (20 meters) there are areas of much larger seaweeds called kelps, which include species such as *Laminaria* and *Alaria*. They often grow in dense clumps referred to as kelp forests, and these are one of the preferred habitats where lobsters hide during the period when they are shedding their old shells. Although invisible from the shore, large kelp forests absorb a great deal of wave energy and contribute to coastal defense. After a major storm, large amounts of kelp are sometimes washed up on the shore. Kelp is harvested for use as a soil conditioner (after the salt has washed out of it), and other substances such as alginates may be extracted from it.

From an ecological perspective, the attached seaweeds (or macrophytes) also serve to oxygenate the water as a by-product of their photosynthesis. But their most vital function may occur after they perish, when they enter the decomposing or detrital food chain that is the life support system for lobsters, crabs, and a host of scavenging invertebrates. Millions of pounds of carbon produced by attached seaweeds are passed up the detrital food chain each year along the coast of the Gulf.

Thus three types of plant communities are found around the shores of the Gulf of Maine. Sheltered places with soft sediment support salt marshes at high-tide level and seagrass beds at lower levels. More exposed places with bedrock or boulders support seaweed communities, with rockweeds and bladderwrack between tidemarks and kelp beds below low tide. In addition to these, there are stretches of shore with sandy beaches that are usually too unstable for the growth of seagrass or seaweed. Their main form of plant production is the growth of microscopic algae on the sand grains, which explains the low fertility of these communities. In many areas, however, stretches of beach may be dotted with rocky outcrops supporting seaweeds, and patches of salt marsh may occur in the shelter of barrier beaches, or where streams or rivers run into the sea. Such areas give rise to the varied and distinctive landscape that makes the Gulf of Maine coast so attractive.

The Special Role of Islands in Nearshore Productivity

Along the Gulf of Maine shoreline, islands distinguish the coast as nowhere else; without them our shores would be both biologically less productive and aesthetically less distinct. Maine, with 4,617 islands listed by the state's Coastal Island Registry, has more islands than any other state or province in the region; more islands, in fact, than anywhere else on the Atlantic coast. This high density of islands sets up conditions for enhanced nearshore productivity.

As the Gulf's water circulates among them, each island helps to mix, oxygenate, and enrich the water. The islands also cause local upwelling of deeper, colder, nutrient-rich water. Tide-induced vertical water currents around islands bring an astonishing abundance of nutrients and marine life up from the sea floor. Tide-driven currents also surge through passages between islands, creating a funnel effect that increases the volume of feed available to filter feeders, as well as those species that prey on the filter feeders. Individually, islands support many wetlands, the runoff from which helps enrich the surrounding waters. These factors help explain the inshore movement of lobsters, crabs, and fish during the spring and summer and the high density of lobster traps around the margins of islands.

Marine ecologists measure productivity by estimating the yearly amount of carbon fixed by plants through photosynthesis. Because plants make up the base of the food chain, these estimates measure what ecologists call *primary production.* The animals that feed on this primary production, the grazers, are called *secondary consumers;* further up the food chain are *tertiary consumers* (predators) that feed on secondary consumers.

For any place on land or in the sea, the amount of plant production is what ultimately controls the abundance of grazers and predators. This is because of what ecologists call the 10 percent rule, which is based on the observation that only 10 percent of the production from one feeding level is passed up the chain and thus available for consumption by the next level—right up to the topmost predators. This relative inefficiency of ecological systems (that is, 90 percent loss with each transfer from one feeding level to the next) explains in part the relative rarity of animals like mountain lions, eagles, or bluefin tuna. In fact, ecological systems are like a complex food web, often likened to a pyramid of life, with predators at the top point, grazers in the middle, and plants residing at the broad base.

The amount of carbon fixed through the photosynthetic process of sunlight interacting with the chlorophyll molecule in plants is reported by ecologists in grams of carbon (gC) per square meter (m^2) per year (yr). Photosynthetic carbon can be measured for both land and sea. Deserts, for example, produce an average of only 70 grams of carbon per square meter per year ($gC/m^2/yr$), lakes and streams 500 $gC/m^2/yr$, tropical rain forests 1,800 $gC/m^2/yr$, and swamps and marshes, the highest in nature, 2,500 $gC/m^2/yr$. Even lands under cultivation produce only 650 to 1,000 $gC/m^2/yr$.

Ecologists at the Bedford Institute of Oceanography in Nova Scotia have been measuring the primary production of seaweeds in the shallows of the Gulf of Maine for more than two decades. Data on nearshore productivity is crucial to understanding how islands scattered along the shore lead to enriched coastal food chains. The estimates are made by studying the growth of seaweeds and by harvesting them. Long-term studies in St. Margaret's Bay, Nova Scotia, show that as much as 1,000 $gC/m^2/yr$ is captured by the seaweeds along the shores there. This compares with estimates of only 125 $gC/m^2/yr$ for the open ocean adjacent to the Gulf and translates to 370 kilograms (815 pounds) of carbon per year for every meter of shoreline.

The coast of the Gulf of Maine is over 7,000 miles in length; thus perhaps as much as 5 or 6 billion pounds of carbon are produced annually merely by seaweeds along the Gulf's coastline. These values are among the highest for coastal productivity recorded anywhere. In part, these high values are due to the large and intricate topography of the shallows around the many islands in the Gulf. Without these islands the nearshore zone would be a much simpler place, without the biological richness we currently observe. Rockweed zones are easily distinguished on satellite images taken at low tide and can be analyzed to pinpoint important intertidal habitats along the Gulf coastline (fig. 7.11).

Perhaps nowhere is the connection between islands and marine productivity more apparent than in the lobster fishery. In the spring of the year, lobsters begin crawling out of their winter feeding grounds toward shore where they will shed their old shells and grow new ones. During this time in their annual cycle, they do not scavenge for food but transform themselves into filter feeders. While their new shells are hardening they are uniquely susceptible to predation and require areas that provide protection. The extensive underwater rockweed and kelp forests surrounding islands in the Gulf provide the necessary cover and rich broth of food. Rockland, Maine, at the edge of Penobscot Bay has long proclaimed itself the "Lobster Capital of Maine," and it is no accident that Maine's largest lobster harvests are found along the section of the coast with the largest number of islands.

The Bay of Fundy Nutrient Pump

The Bay of Fundy is an especially significant part of the Gulf of Maine because of the cumulative effect of the various nutrient sources that contribute to its productivity. It would not be misleading to suggest that the Bay of Fundy acts as a nutrient pump for the northeastern part of the Gulf. In the upper bay, nutrients are cycled out of salt marshes into estuaries and over mudflats; nutrients are almost continuously suspended in the water column by the stirring of the bay's massive tides. Ecologically distinct from the estuaries and salt marshes is the *benthic* or bottom-dwelling community of the mouth of the bay around the West Isles, Grand Manan, and Campobello. Historically the waters around these areas have been among the most productive fishing grounds in the entire Gulf region and have provided feeding grounds for whales, porpoises, and myriad seabirds. The benthic pump of the outer part of the bay appears to be a factor enhancing the nutrient plume of the eastern Maine coastal current (fig. 7.12).

Similar to the funnel effect that operates around islands, benthic pumps exist in areas with numerous shoals and underwater ledges. Essentially, plankton concentrates in these narrow passageways as the water is forced through them. This concentration of food fosters an unusual abundance of filter-feeding invertebrates, which in turn release volumes of larvae and eggs back into the water column, adding even more to the already plentiful plankton base. The West Isles-Campobello-Grand Manan networks of islands, ledges, and shoals are prime areas for funneling to occur. The shores of Grand Manan and the waters off Digby Neck and Briar Island possess an incredibly abundant and diverse population of benthic invertebrates. These include barnacles, sponges, marine worms, and mollusks such as scallops, mussels, and clams.

Although detailed mapping of the currents around the mouth of the Bay of Fundy is yet to be completed, the complex

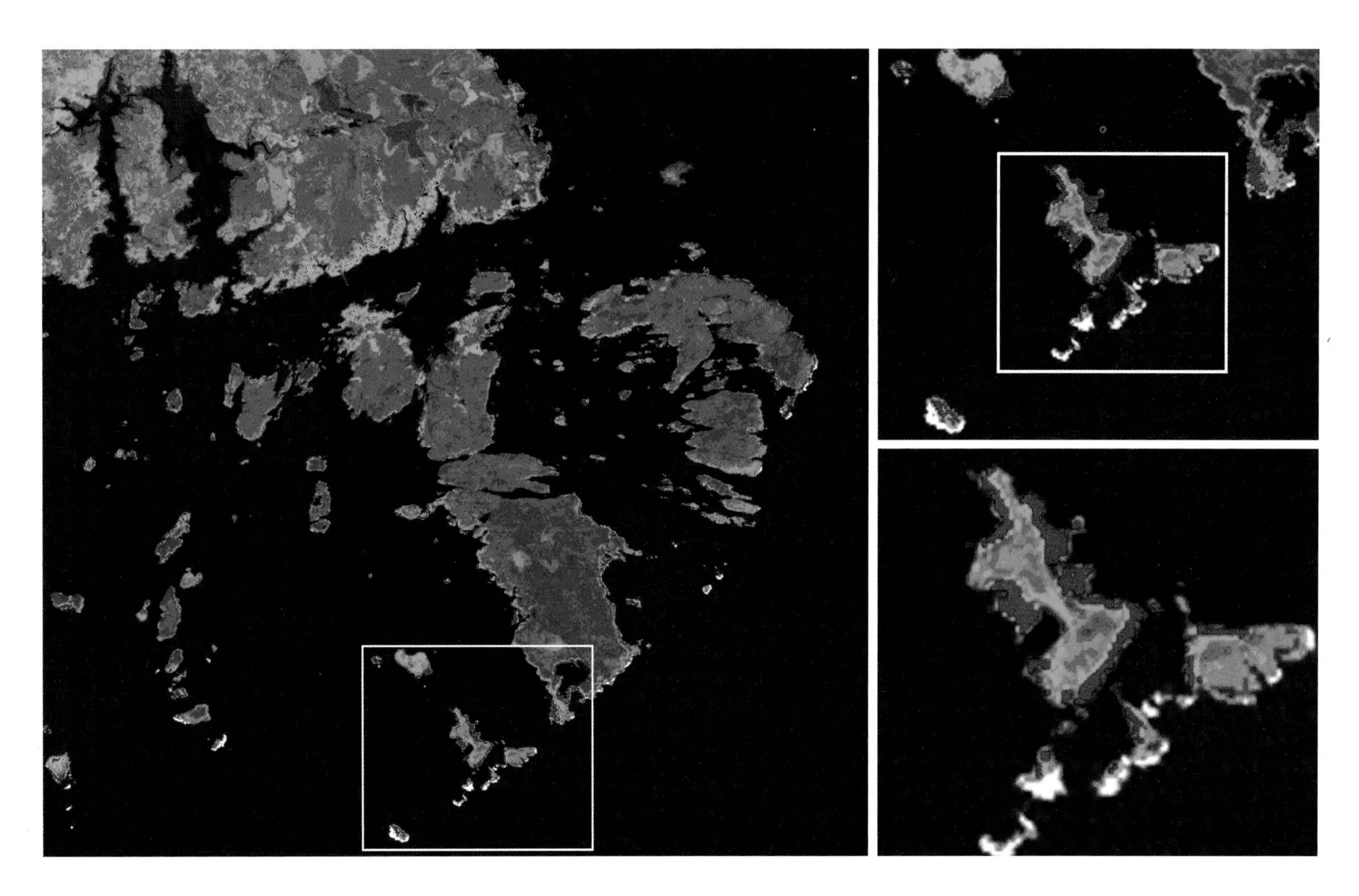

7.11 Intertidal Rockweed Zones

This theme-mapped satellite image shows the distribution of large concentrations of rockweed growth along part of the eastern Maine coast. Rockweed-rich zones are an important source of nutrients for the detrital food chain on which lobsters, urchins, crabs, and other species depend, as well as habitat for juvenile fish. *Credit: R. Podolsky, P. Conkling, Island Institute; GAIA image*

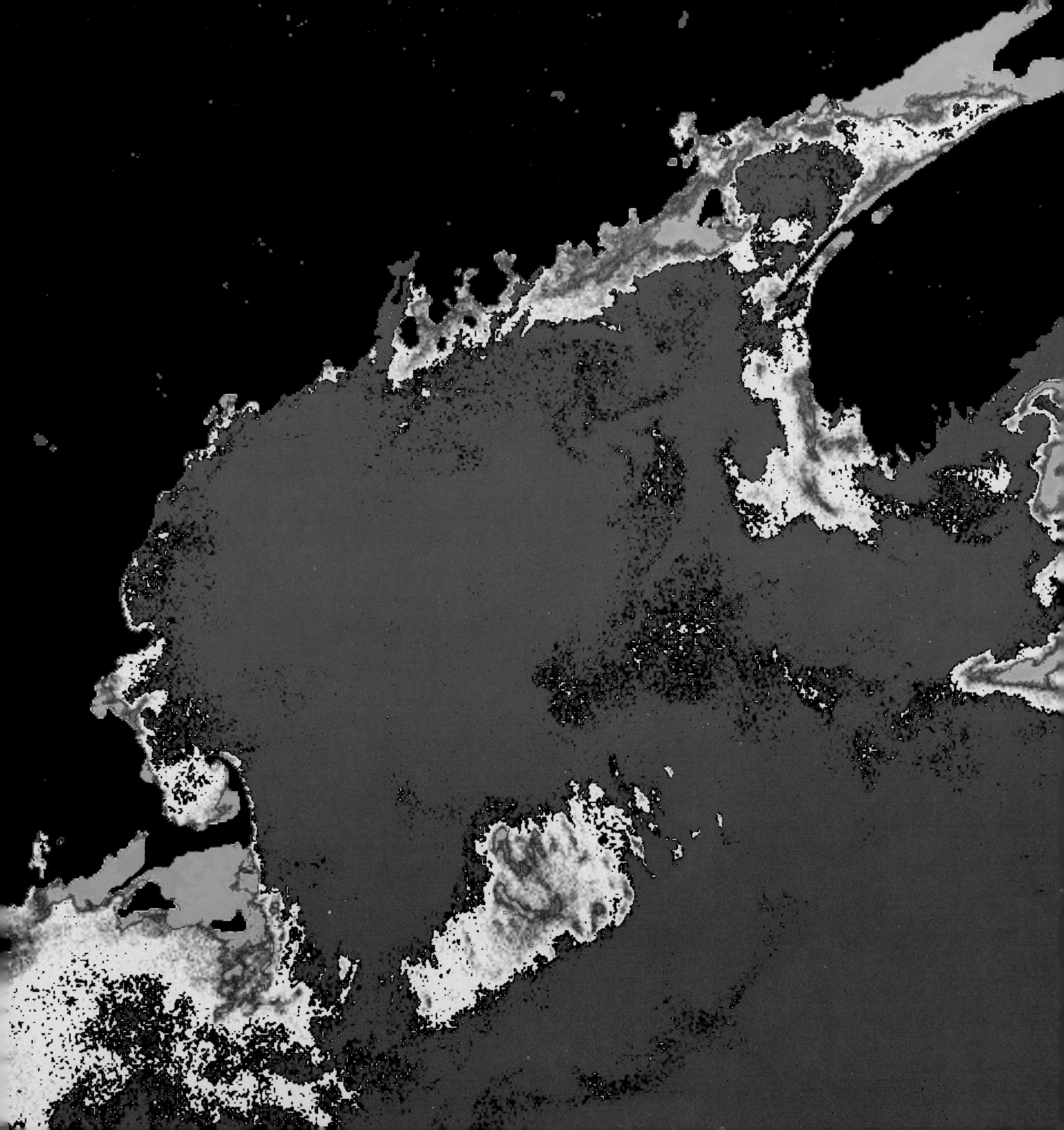

7.12 Bay of Fundy Productivity Pump

This enhanced image shows the extensive zones of high turbidity in the upper Bay of Fundy (in yellow) over its extensive mudflats. The highly enriched phytoplankton zone around Grand Manan Island that trails off southwestward along the eastern Maine coast is also clearly visible (blue-green), and suggests the role that Fundy's currents play in distributing productivity along much of the Maine coast. *Credit: data: Coastal Zone Color Scanner, courtesy of Bigelow Laboratory for Ocean Sciences; image enhancement: R. Podolsky, P. Conkling, Island Institute*

tide-driven currents surging by Grand Manan's enriched subtidal communities also eddy into and around the intricate shorelines of Passamaquoddy and Cobscook bays. The subtidal communities of these areas possess some of the highest marine diversity of cold-water regions thus far identified anywhere in the earth's oceans—a diversity directly related to the interaction of the tides and to nutrient exchange between the inner and outer portions of the Bay of Fundy.

Animal Life in Inshore Waters

All animal life depends ultimately on some source of plant life. Either the animals eat plant material themselves or they prey upon animals that have done so. Yet there are very few animals that graze directly on the large plants of the coastal waters. Instead, the plant material decays in the seawater, breaks into small fragments, and is colonized by large numbers of bacteria. This material, called detritus, drifts about providing food for the myriad clams, crustaceans, and young fish that inhabit the coastal waters.

The rocky intertidal zone of the Gulf of Maine is rich in commercial species of mussels, clams, and marine worms. The total annual harvested value of these species in Maine alone is between $13 and $15 million. These species are nearly ubiquitous along the intertidal rim of the Gulf, but there are some interesting local peculiarities. For example, blood worms are harvested in abundance in eastern Maine but researchers have never found reproducing worms there. This poses two interesting questions: what local circumstances prevent the worms from reaching reproductive condition, and how is the population maintained in the absence of local reproduction? The answer to the latter question may be that larval or juvenile worms may be transported to eastern Maine from the Nova Scotian shore by the counterclockwise circulation of surface waters in that part of the Gulf.

The green urchin is a voracious plant-eater in coastal water. This spherical, spiny animal used to be disliked by lobster fishermen because it was so abundant that it often clogged their lobster traps. When present in small numbers, urchins tend to hide under rocks and feed on drifting plant fragments like other animals. However, from time to time they become very abundant. At such times there are not enough plant fragments to feed all the urchins, so they swarm all over the kelp plants and destroy them completely. Urchins can destroy all the kelp over large areas amounting to thousands of square meters, after which they sit around for years consuming each new plant that attempts to grow. These underwater rocky areas, devoid of seaweed, are described as barren grounds.

On the coast of Nova Scotia, when lobsters are scarce sea urchins become abundant and create barren grounds. It is possible that when lobsters are abundant they help control urchin numbers, and thus protect the kelp beds. Along the Maine coast a $40 million industry has recently and rapidly developed to harvest sea urchins for shipment to Japan. Although no one knows whether the present rate of the urchin harvest is sustainable, it appears likely that kelp beds will be reestablished in many previously barren areas.

Mussels are a normal part of the fauna of rocky shores, especially among rockweeds (fig. 7.13). They feed by filtering phytoplankton and detritus from the water. It is possible to greatly increase production of mussels by suspending them from buoys on ropes or seeding them in controlled densities on the seafloor where water currents are strong. Mussel aquaculture is now a major industry in the region.

From time to time there are explosive growths of particular species of phytoplankton that give off toxic substances (toxic algal blooms). If the mussels feed on these for an extended period they too may become toxic to human consumers. Precautions are therefore taken by the commercial growers to detect the toxic blooms and prevent the marketing of affected mussels. There have been far more reports of toxic blooms in the Gulf of Maine and elsewhere during the past decade than in earlier times. Some think that this is simply because growers and

7.13 Blue Mussels

These bivalved shellfish feed by siphoning water through their systems and filtering out tiny plankton. Because they siphon large volumes of water, they accumulate whatever is in the water column, including contaminants. *Credit: A. Gingert*

scientists are more aware of the problem, but it is a worldwide phenomenon and many scientists think that there has been a real increase. The cause is not fully understood, but increases in the amount of nutrients entering coastal water from agricultural runoff and from sewage are believed to be factors. It is also possible that the toxic species are being spread around the world in the ballast water of ships, which is often discharged into coastal waters.

In places where there are large accumulations of soft sediment and a large tidal range (as at the eastern end of the Gulf of Maine, along the Washington County coast and in the upper Bay of Fundy) we find extensive areas between low and high tidemarks often referred to as clam flats. Researchers have documented that the nutrient source for a significant number of these mudflats originates from the erosion of upland bluffs and unconsolidated glacial deposits along shorelines. Erosion to a landowner can be an energy source for a mudflat (fig. 7.14).

On clam flats, when the tide is in the amount of sediment in the water limits penetration of light. When it is out, there is too much exposure to the atmosphere for most marine species to tolerate. For these reasons, eelgrass beds do not flourish and the flats appear to be nearly devoid of plant life. Instead, the greater abundance of living organisms is found below the surface of the mud.

The commercially important soft-shelled clams are often found in great abundance, along with other kinds of clams. There are very substantial local variations in the reproduction and abundance of clams and mussels, in their growth rates, and in their ages at maturity. A bewildering variety of geological, oceanographic, and meteorological conditions in constantly changing combinations affects these abundances, as does predation by a variety of organisms.

We should expect a good deal of uncertainty in the population estimates of clams or mussels in any given location because these populations are so subject to the vagaries of their larval populations. But what is surprising is the overall decline of clam resources in recent years, particularly throughout eastern Maine where clam populations have declined by as much as 90 percent on thousands of acres of flats. Researchers are trying to determine if there is any connection between the appearance of green algal mats coating the surface of many flats and the decline of clams (fig. 7.15). Although there is much concern over this decline, there is little understanding of or agreement on the causes.

The Edge of the Shore

Salt marshes, estuaries, islands, and intertidal zones all contribute to the enrichment of the Gulf's inshore waters; the larger the salt marsh or the higher the number of islands, the greater the contribution. More and more we see that the edges of shorelines, especially near those areas where fresh and salt water mix, are a sensitive indicator for the health of the entire Gulf of Maine system. Historically, people have always populated the banks of rivers and the coastal edge, where access to transportation is good and the fishing and foraging have always been abundant. The inherent beauty of these natural areas also attracts people. Estuaries are the nursery grounds for many species; salt marshes and marine wetlands are storehouses of food and feeding grounds for migrating flocks; the ever-productive edge attracts mammals both small and large, and the competition there is stiff. Human building and development along this coastal edge is narrowing the margin of success for these ecosystems; whole populations of plants, birds, and animals are being forced to quit them.

Certainly these critical areas are susceptible to the larger geologic and oceanographic forces described in earlier chapters—the configuration of the land, the force of the tides, the effects of weather, and the annual bloom of phytoplankton at the base of the marine food chain. But these areas are also sensitive to what goes on upstream, where freshwater runoff, development, and dams affect the quantity and quality of the water entering the Gulf. The following chapters move our attention ashore and attempt to make the connection between the marine and terrestrial parts of the Gulf of Maine watershed.

7.14 Erosion of Grindle Point, Islesboro

This GAIA student project demonstrates the relationship between erodible shorelines, especially where they have been cleared of vegetation, and intertidal areas where eroded sediments settle out. In the words of the student, "I was offered the chance to help the town of Islesboro reduce the severe erosion problem on Grindle Point. The area sited for help is the shoreline facing Lincolnville at the Islesboro ferry landing. After I researched different plant species, the beach rose, *Rosa rugosa,* appeared to be the best plant to prevent further erosion. The beach rose was chosen for three reasons: it is indigenous to the island of Islesboro, its large thorns deter people from going over the bank, and it has an extensive root system thus reducing further soil erosion. The satellite image of Islesboro helped me identify the vegetative ecology of Grindle Point. Most of the shoreline is gravel and ledge. The area with the worst erosion problem is a man-made park area at the ferry terminal. The erosion is caused not only by the ocean but by the human traffic going down over the embankment." *Credit: Z. Conover, grade 10, Islesboro School; teacher: J. Kerr; courtesy of Gaia Crossroads Project, Bigelow Laboratory for Ocean Sciences*

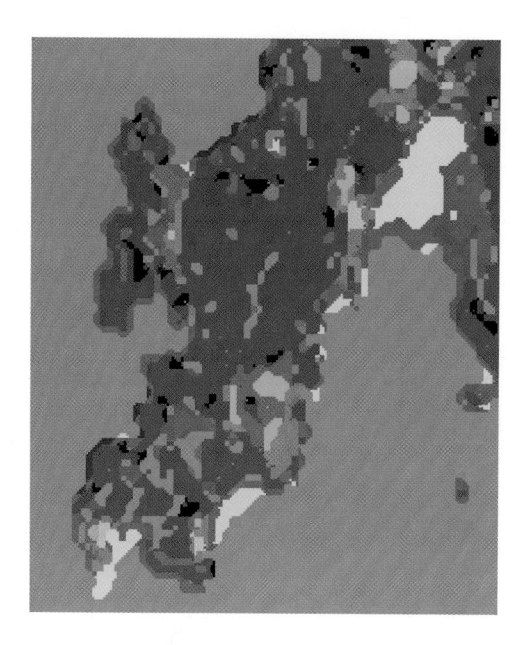

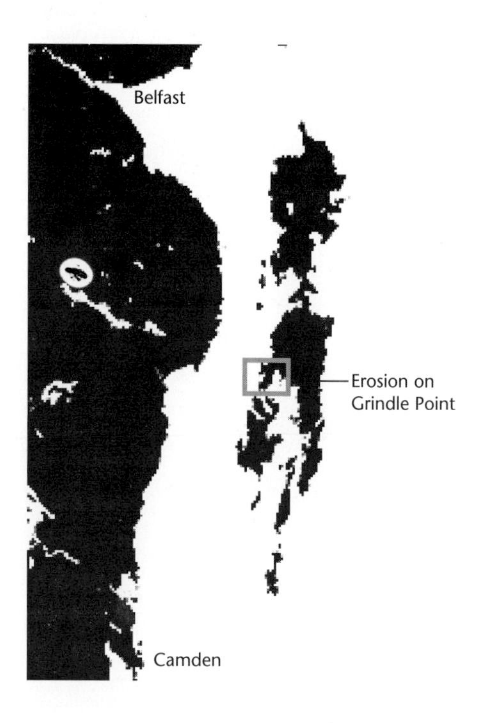

7.15 Algal Ropes on Cobscook Bay Clam Flats

Cobscook Bay clam flats have failed to produce clams for several years. These extensive flats once supplied millions of dollars to the local economy of this isolated area of the Gulf of Maine. No one knows the reason for the decline of the clams, but the flats are now covered by thick mats of green algae. The color aerial photo shows a flat at Weir Cove in Cobscook Bay; the algal mats appear in light green. The second image (following page) is a color infrared aerial in which wetland areas appear in dark shades and the land in red. The third view shows a close-up of the algal mats at ground level. *Credit: B. Beal, University of Maine, Machias*

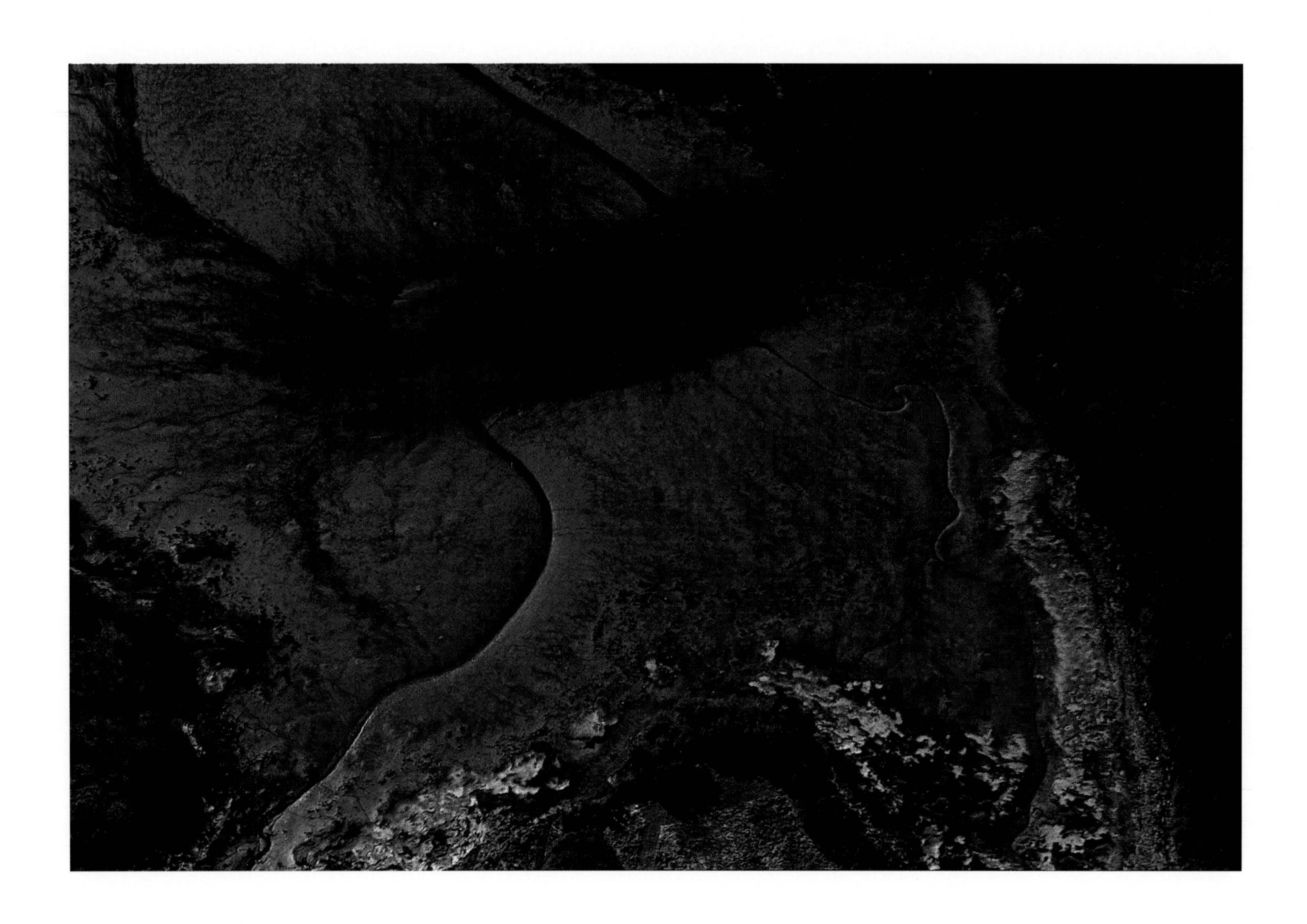

8.1 Top of the Watershed

The upper regions of some watersheds, as in this view of the Presidential Range of the White Mountains of New Hampshire, are located at a ridgeline of a mountainous area. The upper boundary of all watersheds is defined by the place where rainfall begins its long path downslope to the sea. *Credit: A. Gingert*

8

RIVERS, WETLANDS, AND AQUIFERS

HOW WATER MOVES THROUGH THE WATERSHED

Annette S. Naegel, Janice Harvey, Richard Podolsky, and Suzanne Meyer

A *watershed* is a natural boundary shaped by geologic forces and defined by topography. It defines a basin or catchment area that carries the water that is "shed" from the land after rain falls or snow melts (fig. 8.1). Drop by drop, water is channeled into creeks or streams, eventually making its way to large rivers and to the sea. But a watershed is much more than creeks, streams, rivers, and lakes: it is also a highly evolved set of processes that convey, store, distribute, filter, and utilize water, sending it along its course to sustain aquatic and terrestrial life in some obvious and some unseen ways (fig. 8.2).

Watersheds are defined by ridgelines that separate one basin from another. The largest watershed division in the continental United States occurs at the Continental Divide, the highest ridgeline in the United States. Precipitation falling on the eastern slopes of the Rocky Mountains makes its way toward the Mississippi River, while rain or melting snow on the other side of the Divide flows toward the Colorado and other rivers. The waters of the Mississippi eventually reach the Atlantic Ocean via the Gulf of Mexico, while the Colorado's waters reach the Pacific Ocean via the Gulf of California.

The Gulf of Maine Watershed

The Gulf of Maine watershed encompasses part or all of three American states and three Canadian provinces: Massachusetts, New Hampshire, Maine, Nova Scotia, New Brunswick, and a small part of Quebec. Its total area is 69,115 square miles (165,185 square kilometers). Of the six states and provinces, only Quebec does not share the shoreline of the Gulf of Maine, while only Maine is located entirely within the Gulf of Maine watershed.

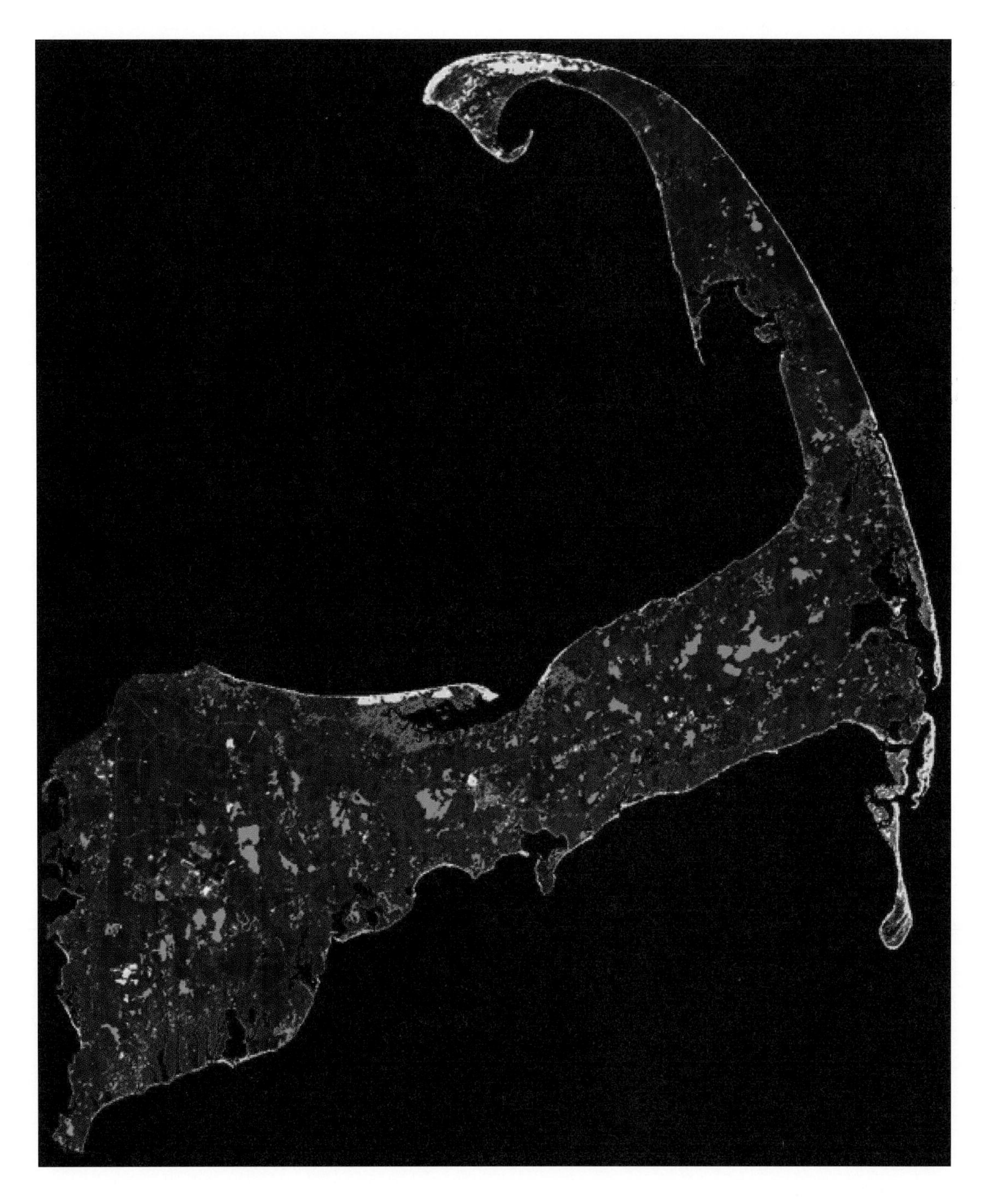

8.2 Surface Waters of Cape Cod

Sand and gravel deposits such as those that underlie most of Cape Cod and the islands are among the most valuable of underground aquifers on earth, based on the purity of the water they store. These aquifers are recharged in part by surface-water lakes, ponds, and kettle holes, shown here in light blue, contrasting with the darker blue marine waters. *Credit: R. Podolsky, P. Conkling, Island Institute; GAIA image*

The land area of the watershed is divided as follows:

State or Province	Area within Watershed: Square Miles	Area within Watershed: Square Km	Portion of Its Land within Watershed (%)	Its Share of Total Watershed (%)
Maine	33,215	79,384	100	48
New Brunswick	15,750	37,642	56	23
Nova Scotia	7,550	18,045	36	11
New Hampshire	6,500	15,535	70	19
Massachusetts	3,400	8,126	41	5
Quebec	2,700	6,453	0.45	4

Figures from Richard Kelly, Jr., Maine State Planning Office

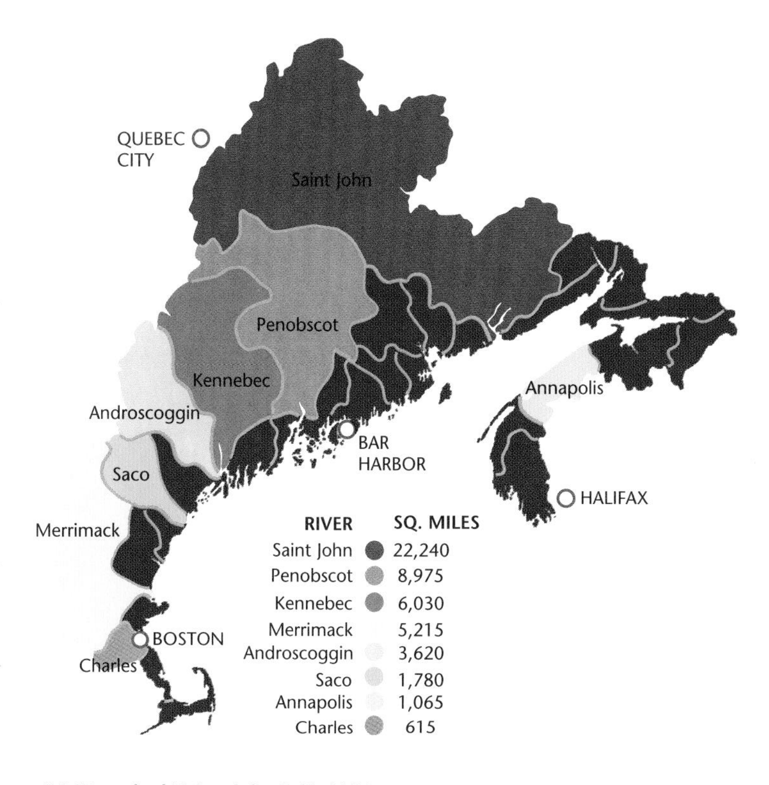

8.3 Watershed Units of the Gulf of Maine

The Gulf of Maine international watershed is made up of a variety of subsheds that in many cases drain substantial areas of land. The Saint John River watershed drains the largest land area, followed by the Penobscot. *Credit: S. Meyer, Island Institute*

The large watershed of the Gulf of Maine is made up of smaller watersheds: streams, ponds, lakes, rivers, and tributaries that drain particular surfaces of the land. The major river drainages are the Merrimack, Saco, Androscoggin, Kennebec, Penobscot, St. Croix, and Saint John rivers (fig. 8.3). The entire watershed contributes an average of 250 billion gallons (950 billion liters) annually into the Gulf of Maine's saltwater habitats.

The Water Cycle

All of the precipitation that falls through the earth's atmosphere is ultimately recycled, with the help of the sun's energy. Only a small portion of the rain, hail, snow, or sleet hitting the ground runs directly into lakes, rivers, or the ocean. Some precipitation is intercepted by leaves, quickly evaporates, and is returned to the air without reaching the earth at all. Most of the water that trees and plants make use of, however, is drawn up through their roots. Water acts as a carrier of dissolved nutrients in the process of photosynthesis, after which it reenters the atmosphere as water vapor in a process called *transpiration.*

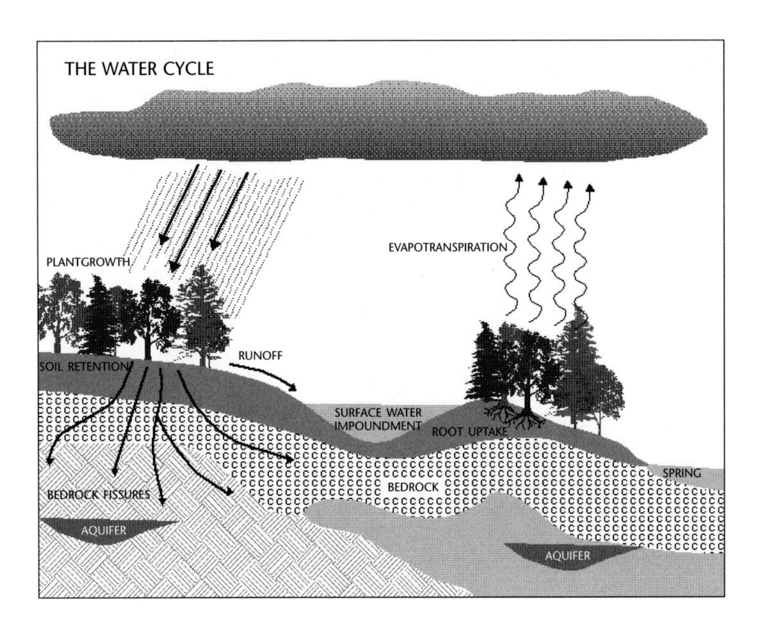

8.4 The Water Cycle

The water cycle begins as precipitation that is intercepted by plant growth, slowing its velocity. Heavy precipitation creates surface runoff that collects in streams, lakes, bogs, and other surface-water impoundments. Another portion of the annual precipitation travels through the soil and enters bedrock through fissures, where it is stored in underground aquifers. This is the most common source of drinking water in the Gulf of Maine region. Part of the water in the soil is also taken in by root systems; the portion not used during photosynthesis, called evapotranspiration, is recycled back into the atmosphere. *Credit: T. Christensen*

Most of the precipitation that falls on a watershed percolates through the soil into bedrock fissures, cracks, and crevices, where it travels below ground as *groundwater*. Much of this water travels downslope until it reaches lakes, streams, and rivers. Even when the ground above is dry, water can be found underground. This subterranean water travels slowly into an *aquifer* (an underground water storage area) until it emerges as a spring, gets pumped out through a well, or seeps into a stream, lake, or ocean.

The water cycle is completed when water is transformed to vapor through evaporation, creating humidity and forming clouds. When atmospheric water condenses, it is reintroduced as precipitation, somewhere else downwind, to continue the water cycle on which regional watersheds depend (fig. 8.4).

Soils and Erosion

Many soil types and landforms influence the course of water flowing through the Gulf of Maine watershed. Much of the region's soil cover was deposited as the last glacier retreated, between 13,000 and 15,000 years ago, and is a mixture of sand, gravel, silt, and clay. Soils play a critical role in distributing water throughout a watershed, because depending on their characteristics they can speed up or slow down water as it moves through the ground. In sand and gravel deposits the soil mix is porous, and water travels quickly and easily. These are known as *permeable* soils. Less permeable layers—clay, for example—act as barriers to water. Following the path of least resistance, water flows readily through sand and gravel but is deflected by areas of clay.

In sand and gravel deposits we find some of the largest and highest-quality aquifers used for drinking water supplies, because these are relatively sterile environments where bacteria have difficulty thriving. Impurities in the water, such as bacteria and other organic matter, are filtered out through sand and gravel beds.

Provided that water is able to percolate through soils, water levels are maintained between precipitation events. Some of the water is held in the soils and made available to plants. The rest constantly recharges groundwater and replenishes aquifers. Even when the ground above may be dry or subject to a prolonged drought, water can be found underground. This subterranean water travels slowly, sometimes over a mile in distance, until it emerges as a spring, is pumped out by a well, or seeps into a stream, lake, or ocean.

Forested headwaters of watersheds are critical to the water cycle because they collect nutrients, absorb energy, slow erosion, and cleanse running water. Erosion and runoff are less likely to occur where there is dense vegetative cover and where high amounts of organic matter, irregular and uncompacted soil surfaces, numerous root holes, and healthy microbic soil activity combine to slow down the flow of water. Where the vegetation has been stripped and soil exposed, in contrast, erosion of mineral soil can happen very quickly and soil may permanently lose its ability to support plants that once grew there.

8.5 Norumbega Fault

Satellite imagery can be used to locate zones of weakness in the ledge or bedrock faults, such as the Norumbega and Sears Island faults shown here. Although a typical well yield is in the range of 2 to 5 gallons per minute, wells drilled into fracture zones may be capable of yielding hundreds of gallons per minute. About 80 percent of the private water supply wells in Maine are drilled into bedrock. *Credit: R. Gerber*

Bedrock Aquifers and Fault Zones

Throughout most of the Gulf of Maine region, drinking water is stored underground in bedrock aquifers—in the tiny spaces or fractures that crisscross rock types. Satellite imagery can be a useful tool in identifying significant bedrock fracture zones along regional *fault* zones, where rock units have slid past one another, intensifying the number of fractures and hence the water storage capabilities. According to Maine hydrogeologist Robert Gerber, water-yielding fault zones are usually linear or curvilinear and can be detected by characteristic changes in tone on infrared satellite imagery. Because water absorbs infrared and near infrared wavelengths, such zones usually appear darker than adjacent areas, reflecting the near-surface concentration of surface drainage and high water table positions that often overlie zones of enhanced deep groundwater flow. Although not all such identified zones ultimately prove to be above average in well yield, studies in the midcoast area of Maine suggest that about two-thirds of the high-yield wells can be associated with fault zones mapped from satellite imagery or aerial photography. The Norumbega Fault zone seems to be a particularly productive fracture system (fig. 8.5).

Fracture tracing, as the process of mapping bedrock fracture zones from high-altitude imagery is called, is also useful in other aspects of environmental planning. It can be used to map the susceptibility of waste disposal sites to leakage and hence their likelihood to contaminate aquifers. It can identify potential zones of saltwater intrusion in aquifers, since high-yield zones connected to the ocean are more susceptible to intrusion. Fracture tracing enhances our capability to plan for orderly development and solve some large-scale environmental problems.

The Special Case of Island Aquifers

The islands in the Gulf of Maine each have their own watershed systems that capture, store, and use water that falls upon their surfaces. Precipitation replenishes island groundwater in the same way as it replenishes mainland water supplies. On an island, however, that water supply is self-contained. Island water all comes from precipitation, and there is little distinction between groundwater and surface water—it is all connected. For these reasons, islands in the United States have earned the designation of "sole source aquifer" from the United States Environmental Protection Agency.

Surface and subsurface fresh water on an island floats on the salt water that permeates the rock below. Salt water is 1.025 times heavier than fresh water, and in a cross section the fresh water appears as a lens, deep and rounded in the center and tapering to a sliver where it meets the sea at the island's edge (fig. 8.6). Because of this phenomenon, it is possible to pump

8.6 Island Aquifer

Adequate supplies of water exist as a freshwater "lens" because fresh water is less dense than salt water. This resource extends down to a depth 40 times the height of the water table (*h*) under most islands larger than 5 acres in the Gulf of Maine. For example, if the water table is one foot above sea level, a 40-foot-deep reservoir will, on average, be located beneath. Difficulties occur, however, when wells drilled too near the shoreline begin pumping significant amounts of fresh water, which can lead to saltwater intrusion. Freshwater resources on islands are special cases because they are "sole-source aquifers," replenished only through annual precipitation and unconnected to other freshwater supplies, unlike mainland aquifers. If island aquifers become contaminated, it is usually impossible to repair the damage. *Credit: T. Christensen*

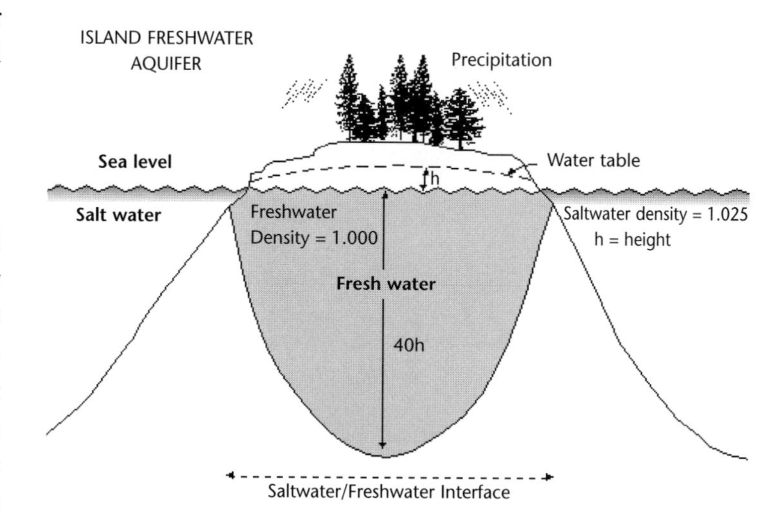

drinkable water out of the ground on an island, even though the island is surrounded by the sea.

If too much water is pumped too fast, salt water can be drawn into an island or coastal well. Seasonal salinity is a problem known to landowners all along the coast. If not enough penetrable surface area is available to recharge the groundwater, water supplies can dry up. If the water reserves become contaminated, fresh water may not be available; malfunctioning septic systems or leaking oil storage tanks are a high risk to groundwater supplies. The underground lens of fresh water only holds so much, and replacing salty, depleted, or polluted water can be economically devastating, if possible at all. The damage to contaminated groundwater is in a sense irreversible, since it could take generations to restore the aquifer. Understanding the link between island water supplies and the watershed is vital to preventing such catastrophes.

The Role of Wetlands

Within the course of a regional watershed, there are many different surface water features falling into the broad classification of wetlands: fast rivers, small streams, meandering oxbows, large lakes, shallow ponds, coastal salt marshes, and forested wetlands (fig. 8.7). Each type is a specific habitat for certain plant and animal species. Wetlands are the transitional environments between terrestrial and aquatic systems, and are highly productive ecosystems. Wetlands are nutrient sinks, the basis of complex food webs that support valuable populations of furbearers, waterfowl, and other wildlife. Wetlands also play a critical role in retaining and controlling floodwaters, trapping sediments buffering shorelines from erosion, and recharging groundwater supplies. Wetlands act as recharge areas and filters for groundwater reservoirs that provide three-quarters of the drinking water in the United States. Forested areas along headwater streams maintain cool water temperatures, supply organic matter, provide woody debris, and shelter many kinds of wildlife. The rivers that connect upland streams and wetlands with the Gulf of Maine help to purify water by processing and filtering nutrients and other pollutants. Finally, wetlands provide water for recreation and are places of scenic beauty. Wetlands are one of the most fragile and important landscape components in North America and throughout the world.

Since European settlement, two-thirds of the Atlantic coastal marshes have been lost, mainly to agricultural expansion, including nearly half of the wetlands within the Gulf of Maine watershed. Development has also put pressure on wetlands; the expansion of industries, cities, and towns in the coastal zone and along rivers has led to draining and filling. Pollution from all sources has contaminated wetlands and the natural communities that thrive within them.

Following water through the landscape is difficult because it takes so many different guises as it travels from freshwater streams through swamps and swales to brackish ponds and into salt marshes. Furthermore, serious wetlands study is a relatively new science, going back less than two decades. One problem is that a large number of Gulf of Maine wetlands are literally invisible to the agencies that are supposed to protect them. In the United States, for example, wetland patches smaller than three acres are below the minimum size easily identified in standard aerial photography; consequently, they are not mapped by the agencies charged with their protection. Wetland patches larger than three acres are identified by officials at the U.S. Fish and Wildlife Service and are included in the National Wetlands Inventory, copies of which are circulated to state and local agencies across the country.

Satellite image analysis and geographic information systems will be increasingly important tools in quantifiying the study of wetlands. Landscapes can now be measured using technology that allows the analysis of smaller parcel sizes of important features. Recently school children in a variety of towns, including South Portland, Thomaston, and Wells, have used satellite images to help map town wetlands. Such technology

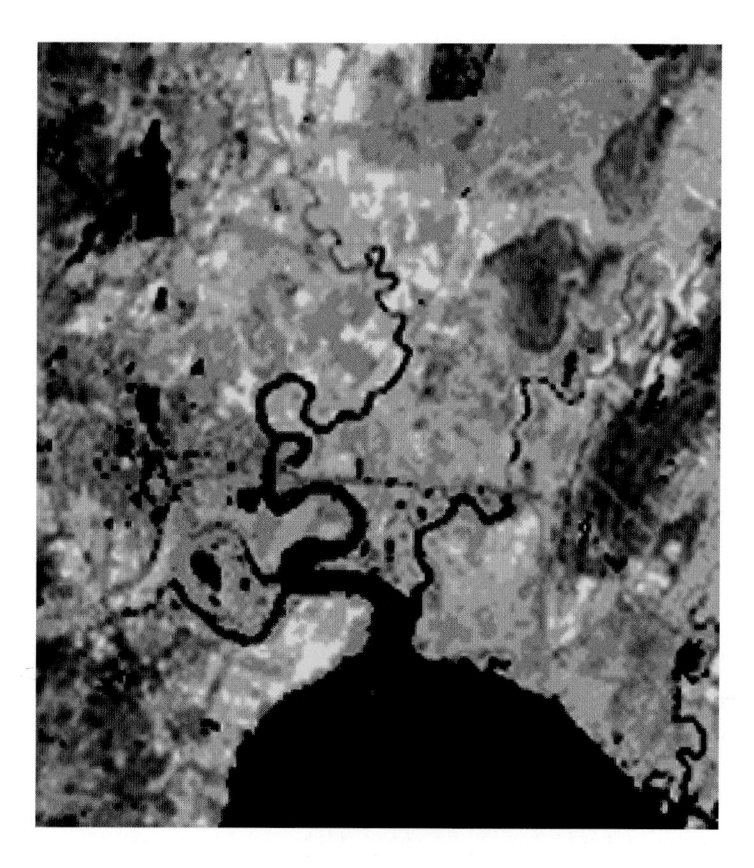

8.7 Oxbow Meander

Oxbows are unusual water features that form where the gradient of a river or stream is slight. The watercourse winds around creating "lazy" turns, as shown here from a marsh system from New Brunswick. Occasionally these meanders are cut off from the main current of the stream to create oxbow lakes, which can also be seen in this image. *Credit: S. Meyer, T. Ongaro, Island Institute; GAIA image*

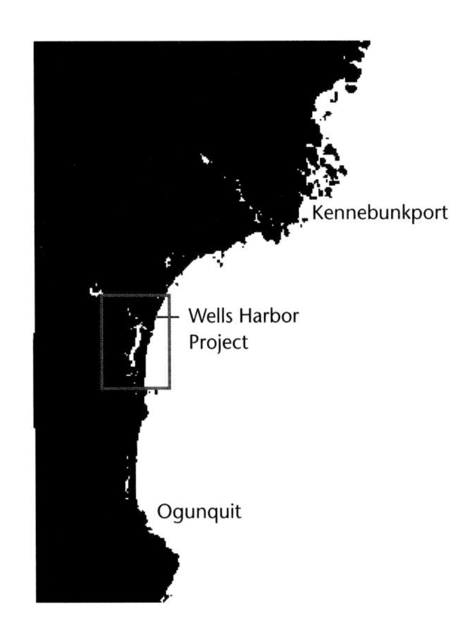

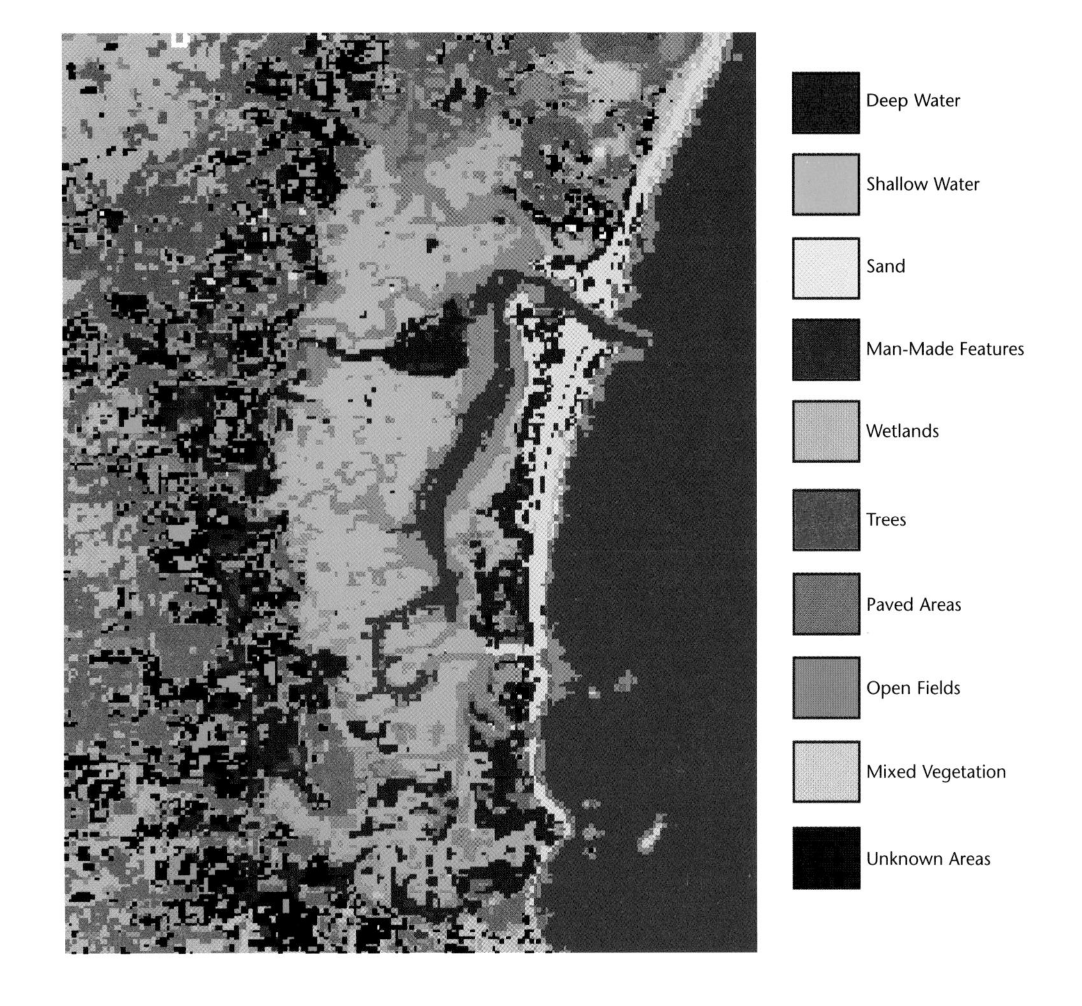

8.8 Wells Harbor Wetlands

This GAIA student project shows developed areas and wetland areas around Wells Harbor, Maine. According to the students, "We chose this area because not many of us were real familiar with the area and thought it would be a challenge. It took us a while to figure out what all the features on the [satellite] pictures we received were, and often we had to go explore the region to find out what they were. This project took us about three months. From this experience we learned more about our town and what it is made of. And we learned many helpful computer skills. We liked this project because it was both fun and educational." *Credit: K. Heon, C. Jarochy, S. Pinard, grade 10, Wells High School; teacher: L. Ryan; courtesy Gaia Crossroads Project, Bigelow Laboratory for Ocean Sciences*

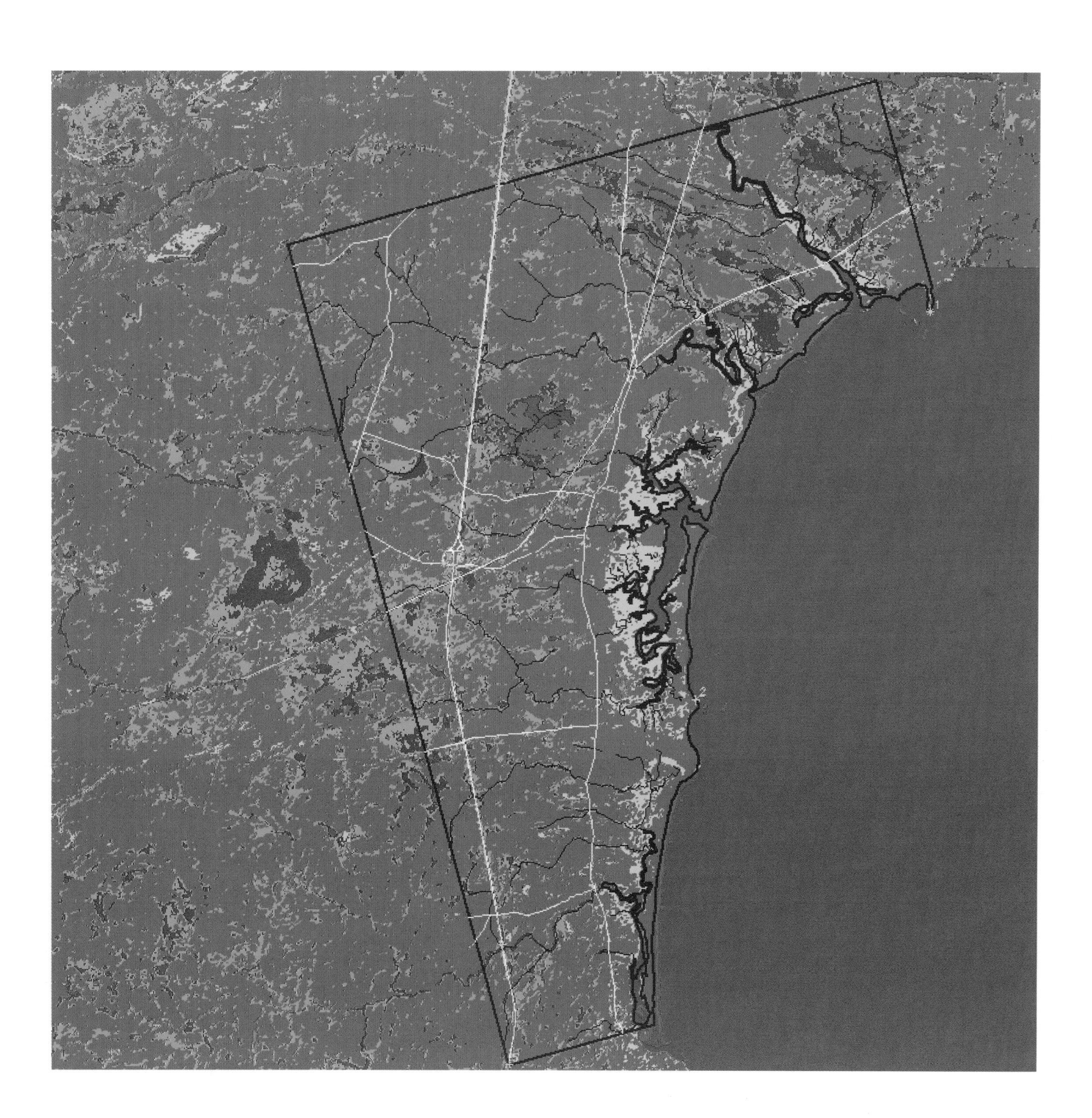

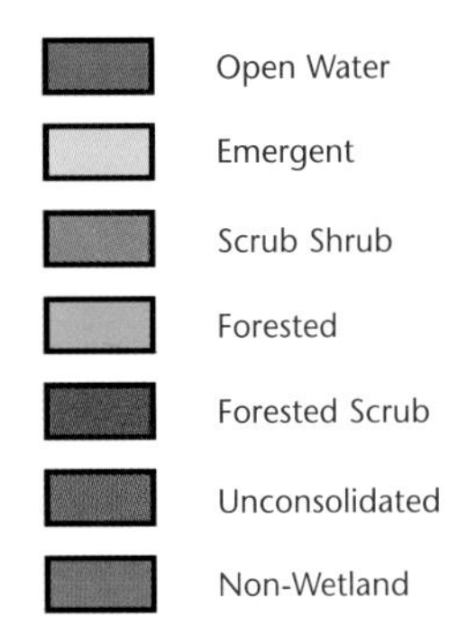

8.9 Patch Analysis of Southern Maine Wetlands

Patch analysis of the area shown within the USGS quadrangle of Wells (outlined in blue) indicated a total of 2,579 discrete wetland patches, the vast majority of which fall into parcels of less than 10 acres. Of the 10,387 wetland acres analyzed, 15 percent are in parcels of less than five acres and 20 percent are in patches of less than 10 acres. This analysis indicates that as much as one-fifth of the wetland acres in coastal Maine may be too small to be afforded legal protection, even though they constitute a large fraction of the total. *Credit: R. Podolsky, P. Conkling, Island Institute; GAIA image*

was also used by the Island Institute to map and analyze wetlands in a 120,000-acre area of south coastal Maine encompassing the towns of Kennebunk, Wells, and Ogunquit (fig. 8.8). In this image, 23,000 acres, or 19 percent of the total area covered, are in one of five wetland cover types. The Island Institute conducted a detailed parcel or patch analysis of wetlands in a 54,531-acre subarea (fig. 8.9). Forested and forested scrub/shrub wetlands were the most common wetlands in the study area, accounting for 60 percent of the wetlands.

Watersheds and Land Use

The character of the water that reaches the Gulf of Maine is dependent largely on two factors: different land uses and the amount and type of waste discharged into streams and rivers. The removal of forests (fig. 8.10), the use of agricultural chemicals, industrial and urban discharges into streams, and the impoundment of flowing waters all affect human and natural communities downstream as well as the Gulf (fig. 8.11).

The cleaner the streamflow that leaves the headwaters of the Gulf of Maine watershed, the more valuable it is downstream to wildlife, plants, people, and industry. However, along the coast of Maine and in much of the Bay of Fundy region, disease-causing bacteria from untreated human wastes have closed thousands of acres of clam flats to harvesting. Mussel harvesting is permanently closed in the Bay of Fundy. New Brunswick fishermen confront pollution every time they sail into the plume of the Saint John River where it spills out into the bay—they smell it. Weir fishermen and scallopers have been closed out of the upper L'Etang estuary. Pulp mill pollution robbed those waters of oxygen, killing or driving out every fish.

Point Sources of Pollution

Pulp mills, paper mills, and tanneries have historically been located at the edges of rivers because their processes require large

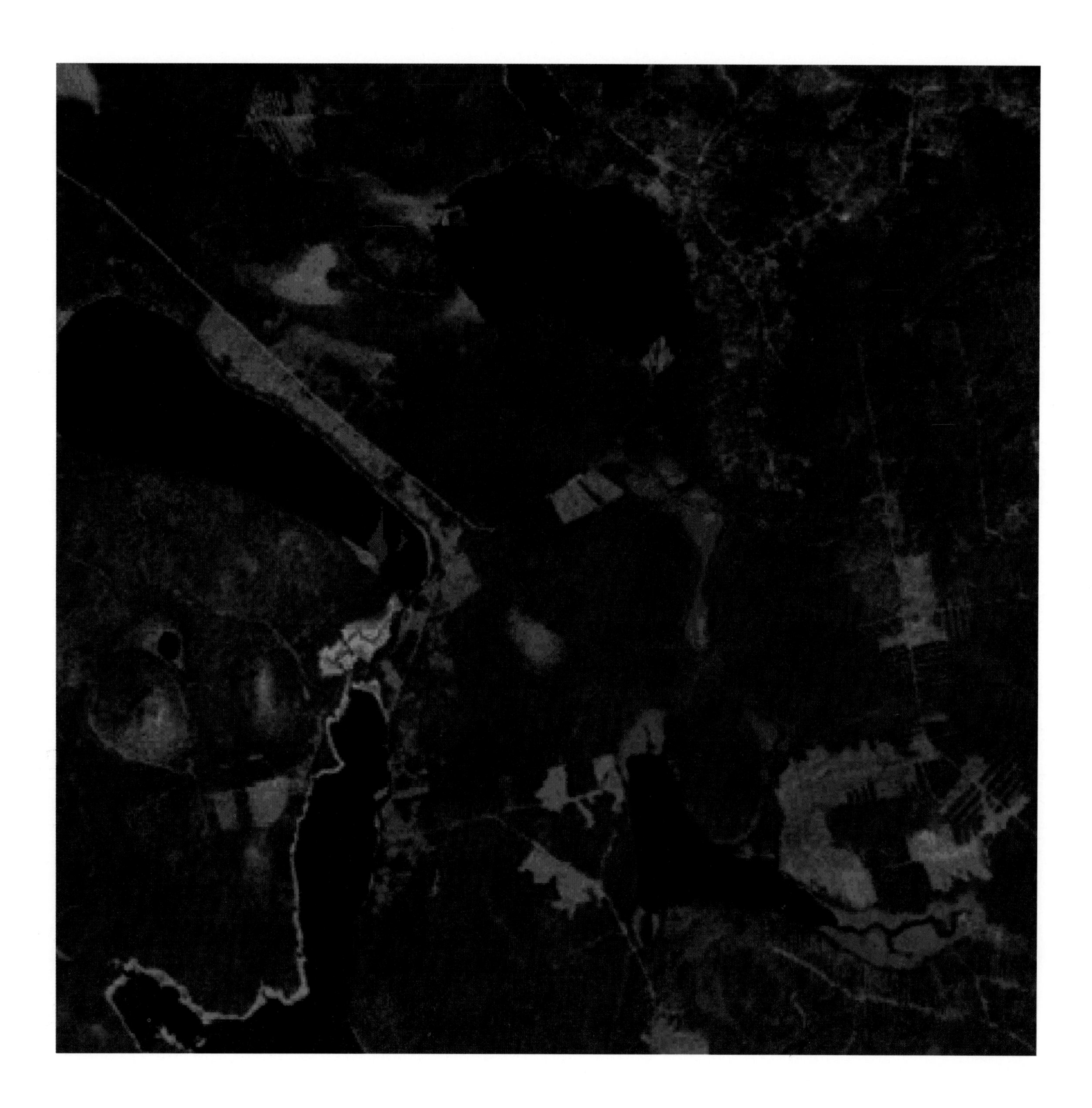

8.10 Lakeshore Cutting and Stream Silt
(left)

The pattern of large-scale clear-cuts is evident in this close-up of a 1983 Landsat image during the height of the salvage cutting for spruce budworm in northern Maine. Clear-cuts are shown in lighter shades; note what appears to be a sediment plume (in purple), which may be the result of heavy cutting near the shore of Umbazookas Lake in Township 6 Range 13. Note also the "beauty strip," the uncut stand of trees along the northern shore of Umbazookas. Umbazookas drains south into the West Branch of the Penobscot while Mud Pond (near the top of the image) drains north into Chamberlain Lake and the Allagash Wilderness Waterway. The winter road that runs between Umbazookas and Mud Pond runs through low, wet ground between these large drainages. *Credit: P. Conkling, T. Ongaro, Island Institute; GAIA image*

8.11 Allagash Wilderness Waterway Beauty Strip
(right)

The "beauty strip" shown in this aerial photo of the Allagash Wilderness Waterway is the thin stand of trees that separates the waterway from industrial forest landscapes a short distance away. *Credit: C. Ayres*

amounts of water. In the case of paper mills, logs were driven down from the headwaters to these processing centers. In the making of pulp and the bleaching of paper, water is drawn from upstream surface water, and treated wastes are discharged downstream, becoming pollution from *point sources* (figs. 8.12, 8.13). Dioxin, a toxic by-product of the chlorine bleaching process, ends up in Gulf of Maine rivers in sufficient quantities to justify fish advisory warnings on 235 miles of Maine rivers, and a warning that pregnant women not consume the tomalley (liver) of lobsters caught along the Maine coast (fig. 8.14). Paper-making processes in New Brunswick are producing similar results. Governments on both sides of the border, in the face of less-than-certain scientific evidence and industry pressure, have been reluctant to ban dioxin-producing processes outright. Scientists are increasingly concerned about the effects of exposure to low levels of chlorine-containing chemicals on the development of living creatures, including humans. In Maine, the largest volumes of chlorine that are discharged into fresh water are associated with the old Great Northern pulp mill in Millinocket, which legally discharges 14,000 pounds of chlorine into the Penobscot River annually.

In Saint John, New Brunswick, the largest volume of industrial waste flowing into the harbor comes from three pulp mills that discharge a total of 130,000 cubic meters of effluent per day. Much of this volume consists of woody wastes. When these settle on the bottom, they physically smother the harbor bottom, damaging habitat. Once they begin to decompose, the wastes rob the water of oxygen. The natural chemicals concentrated in this effluent are often lethal to marine life, and cause some types of fish to avoid the area altogether.

At the upper limit of Fundy's tidal reach, the Petitcodiac River flows past Moncton, New Brunswick's second largest city. It was once known for its spectacular tidal bore, one of the best shad fisheries on the Atlantic seaboard, and a healthy Atlantic salmon run that populated a vast network of streams and tributaries well above Moncton. Building a causeway across

8.12 Industrial Sources of Pollution

The U.S. side of the border contributes effluent from over twice as many industrial sources (88) as the Canadian (42). *Credit: S. Meyer, Island Institute; data courtesy of Gulf of Maine Council on the Marine Environment*

8.13 Point Sources of Pollution in the Gulf of Maine Watershed

This map shows 279 major and significant minor point sources of pollution in the Gulf of Maine for which latitude and longitude coordinates were available. Not surprisingly, these discharges are concentrated around the watershed's major cities, including Boston, Portsmouth, Portland, and Saint John. Industrial and municipal wastewaters also enter the Gulf of Maine via the major rivers such as the Merrimack, Kennebec, Androscoggin, Penobscot, and Saint John, where such facilities line the shores. The United States has over ten times as many municipal sewage treatment plants (137) as Canada (12) emptying into the region. Many of Canada's smaller towns and cities discharge untreated or partially treated wastes into Canadian waters. Information is for the period 1991–1993. *Credit: S. Meyer, Island Institute; data courtesy of NOAA Geocoast Facility*

8.14 Dioxin Monitoring Sites

In 1984 the United States Environmental Protection Agency began monitoring dioxin in the Gulf of Maine region. They chose a supposedly pristine control site on the Androscoggin River in Maine—which turned out to have high levels of dioxin from upstream paper mills. Due to the importance of lobster to the local population and economy, the Maine Department of Environmental Protection began its own dioxin monitoring program using lobsters in 1992. Because the lobsters' tomalley functions like the liver in other species, removing chemicals and toxins from their systems, researchers anticipated finding high levels of contaminants in the tomalley. To their surprise, they also found high concentrations of TCDD, the most toxic form of 200 types of dioxin known. The meat from the same lobster, however, was perfectly safe. This graphic shows the four sites on four rivers where the lobsters for the study were taken. On February 4, 1994, a consumption advisory was issued, warning against the eating of lobster tomalley. *Credit: S. Meyer, Island Institute; data supplied by B. Mower, Maine Department of Environmental Protection*

the Petitcodiac to Riverview ended all that. Until recently Moncton flushed 109,100 cubic meters of raw sewage into the river per day, including human waste and the cocktail of chemicals that are poured down the drains in households, businesses, hospitals, and universities. In 1994 Moncton's sewage began to receive treatment to screen and settle out solids, but this will do little to reduce the biological oxygen demand and the levels of toxins in the sewage effluent.

The combination of uncontrolled pulp mill pollution and the reduction in water flow caused by another causeway brought about the demise of the L'Etang River estuary as a productive fishing area. Nine years after the Lake Utopia pulp mill opened in 1971, eight kilometers of the 14-kilometer estuary were badly polluted and the estuary bottom was smothered with wood waste. Of 50 bottom-dwelling marine species previously found there, 46 were eliminated. Oxygen in estuary waters was severely depleted for three kilometers below the causeway. The scallop clam and herring weir fisheries were destroyed, and the rotting wood wastes on the bottom expelled a strong sulfur odor like rotten eggs throughout the area.

Non-Point Sources of Pollution

In downtown centers, malls, airports, and industrial parks across the Gulf of Maine region, where impermeable surfaces—roads, large parking lots, and extensive roofs—cover natural recharge areas, precipitation cannot percolate through soils to replenish groundwater supplies. Here, runoff increases dramatically, resulting in high-velocity water flows and increased erosion; here, too, water picks up all kinds of oil, lead, and other contaminants, which are collectively referred to as *non-point-source pollution.* If contaminants are spilled or dumped on the ground or flushed into septic systems, they may render groundwater supplies unfit for consumption. Proper siting of land use activities within recharge areas and district water supplies helps to ensure abundant and safe groundwater supplies.

Agricultural practices can significantly affect water quality in nearby streams. Pesticides, herbicides, and fungicides sprayed on or injected into agricultural lands work their way into soils, groundwater, and nearby streams, affecting both soil bacteria and aquatic organisms. Recent reports from the Atlantic Sea Run Salmon Commission have detected in several salmon spawning rivers minute quantities of an agricultural herbicide, Velpar, that is extensively sprayed on blueberry lands in eastern Maine. Excessive animal waste from livestock can add excessive nitrogen to streams, fertilizing aquatic plant production and reducing available oxygen for other aquatic organisms. Domestic livestock can trample the banks of freshwater streams, causing erosion and the destruction of the vegetative buffers.

Flowing waters can be degraded by the cumulative effects of deforestation, agricultural runoff, urban and suburban runoff, and more. By the time they reach the coastal zone, the steeper gradients flatten out, the channels widen, the water slows down, temperatures increase, and oxygen and nutrient loads decline. Here moving waters settle into the floodplains and marshlands before being flushed out to the Gulf of Maine (fig. 8.15).

Dams

The tremendous quantities of fresh water that flow into our rivers provided the first cheap source of power for industry across the entire Gulf of Maine region. Many major industrial areas were located in a broad arc where rivers cascading out of highlands went over a falls. Towns like Lowell and Lawrence on the Merrimack, Lewiston and Auburn on the Androscoggin, Augusta on the Kennebec, Orono and Old Town on the Penobscot, and countless others owe their establishment to their location where fast-flowing rivers could be dammed.

Today, although comparatively little of the region's electricity is derived from hydropower, most of the dams that were originally constructed to harness the energy of free-flowing

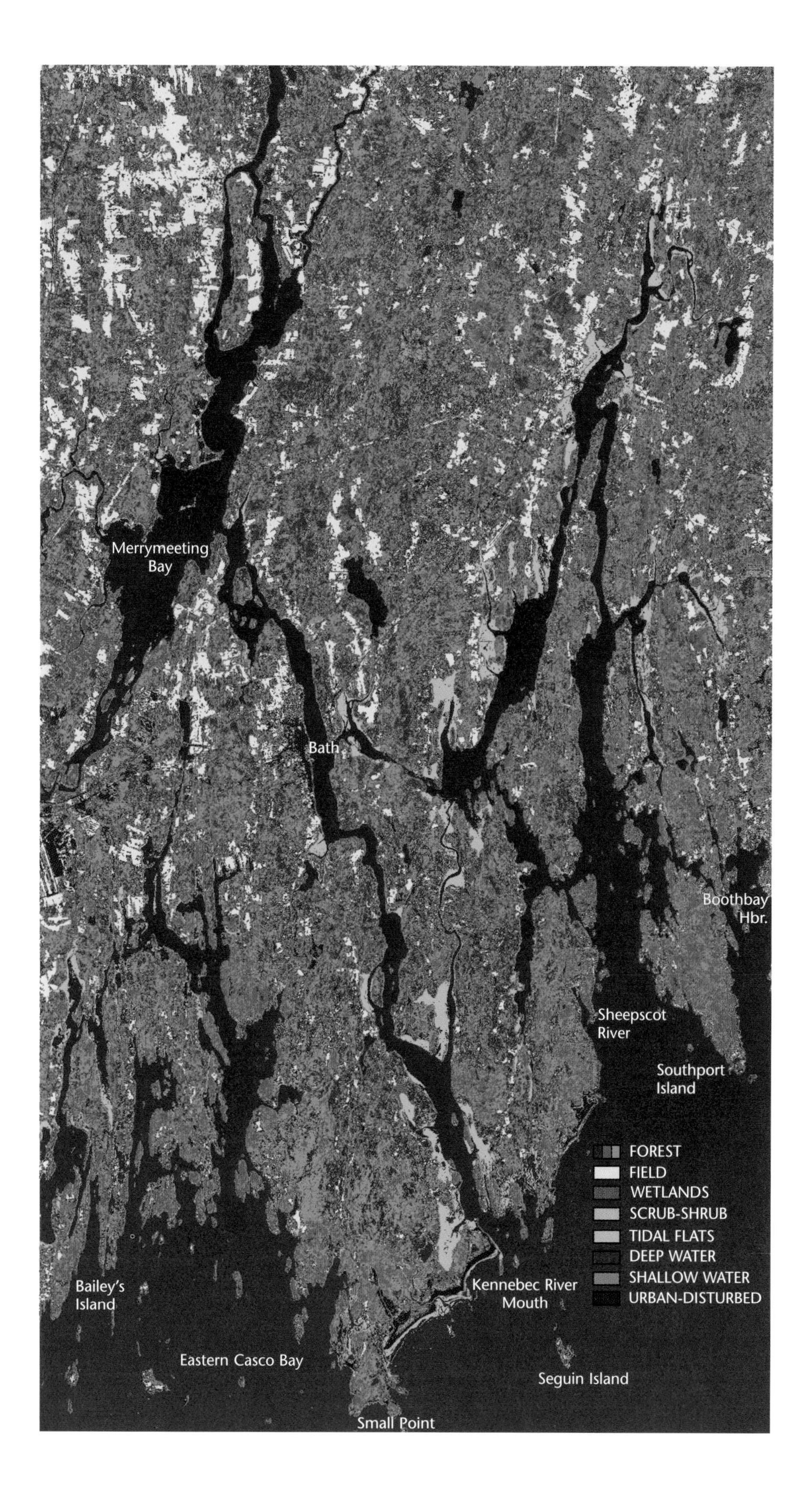

8.15 Non-Point Runoff Surfaces in the Kennebec Estuary

One of the principal advantages of satellite imagery is the ability to conduct queries concerning large land areas, or to look comprehensively at an area as geographically complex as the sharply indented coastline of midcoast Maine. This SPOT satellite image analyzes 316,000 acres, of which 100,000 are marine. The investigators proposed looking at the relationship between fields (yellow) and marine wetlands (lighter blue). Researchers hypothesize that saltwater marshes, intertidal zones, and clam flats adjacent to fields and developed areas will be more likely to sequester chemical pollution that runs off agricultural fields, lawns, and paved areas. To test this theory, this satellite information could be used to select intertidal areas and compare water quality at different sites. *Credit: R. Podolsky, P. Conkling, Island Institute; GAIA image*

rivers are still in place. No one really knows the total cumulative impact these rivers have had on nearshore environments where they meet the sea. We do know that great quantities of nutrient-rich sediment build up behind dams and that this important source of energy is no longer available to the food chains downstream. More significant symbolically, if not actually, is the effect such dams have had on *anadromous* fisheries, especially on the Atlantic salmon which was once the most prized food item in the region.

In 1993, the Minister of Fisheries and Oceans in Canada closed all rivers flowing into the Bay of Fundy to recreational and native salmon fishing. The Atlantic salmon populations in those rivers are the lowest ever recorded. Those salmon that are returning to their spawning rivers will produce only half enough eggs to support a healthy population. This collapse occurred in spite of the fact that the Fundy commercial salmon fishery, for all intents and purposes, was essentially closed in the early 1970s. While scientists are reluctant to provide explanations, a brief look at Fundy's major salmon rivers will shed some light on at least part of the problem. Virtually all of them have or have had dams blocking both the flow of water and the movement of migrating fish.

In the outer bay, the St. Croix River is the artery that links a vast network of lakes and rivers on the New Brunswick-Maine border. Designated as a Canadian Heritage River, it distinguished itself early, first by its rich fisheries, second by the dozens of sawmills constructed on its banks. With the sawmills came dams, most of which made no provisions for fish, contrary to provincial law in the mid-1800s. The decline of the timber trade gave rise to the pulp and paper industry. The dams were turned over to Georgia Pacific Corporation, which uses them to control water levels on the river. Unfortunately, water levels often change dramatically, creating problems for species in the river. Only recently have improvements been made in the fishways on the dams to enhance fish passage. Along much of the river, rotting logs from the log-driving era blanket the bottom, inhibiting gravel-loving fish from spawning.

In Maine and Massachusetts, the situation with regard to the Atlantic salmon is much the same (fig. 8.16). Despite millions of dollars of federal funds spent in past decades to raise and release small salmon that might be caught on their return to natal rivers, the stocks everywhere have declined to alarmingly low levels. The Penobscot River run of salmon, once Maine's largest, has dwindled to a few hundred returning adults.

Besides interfering with fish migrations, dams on rivers create additional problems. At the western boundary of Fundy National Park, for example, the shoreline changes from hard rock to soft sandstone. From here to the head of the Bay of Fundy, vast mudflats are exposed to higher and higher tides as the bay narrows. All the rivers flowing into this upper zone at Shepody Bay have been dammed. Shorelines and banks erode easily and sediments are suspended in the water column, creating a murky, muddy aquatic world. When the strong flushing action of the tides and river flow is slowed by dams, sediments build up, narrowing the river channels and changing the entire system. Reduced flow leads to reduced mixing of fresh and salt water in bays. Less tidal flushing in dammed rivers and coves leads to lower water quality and lower biological diversity. The small-toothed Salisbury pilot whales which used to follow prey into the Petitcodiac River estuary do so no longer. The endangered dwarf mussel is no longer found there. The range of species once present in the freshwater zones of these rivers is in decline. The impacts on ecosystems above these dams have not been studied.

Watershed Consciousness

After decades of intensive coastal development, we are beginning to appreciate the importance of the natural systems we inhabit and the impact we have on them. New land use models are beginning to be implemented that recognize the viability of whole watersheds, and our role within them. Around the Gulf of Maine some groups have recently become active in reshaping our

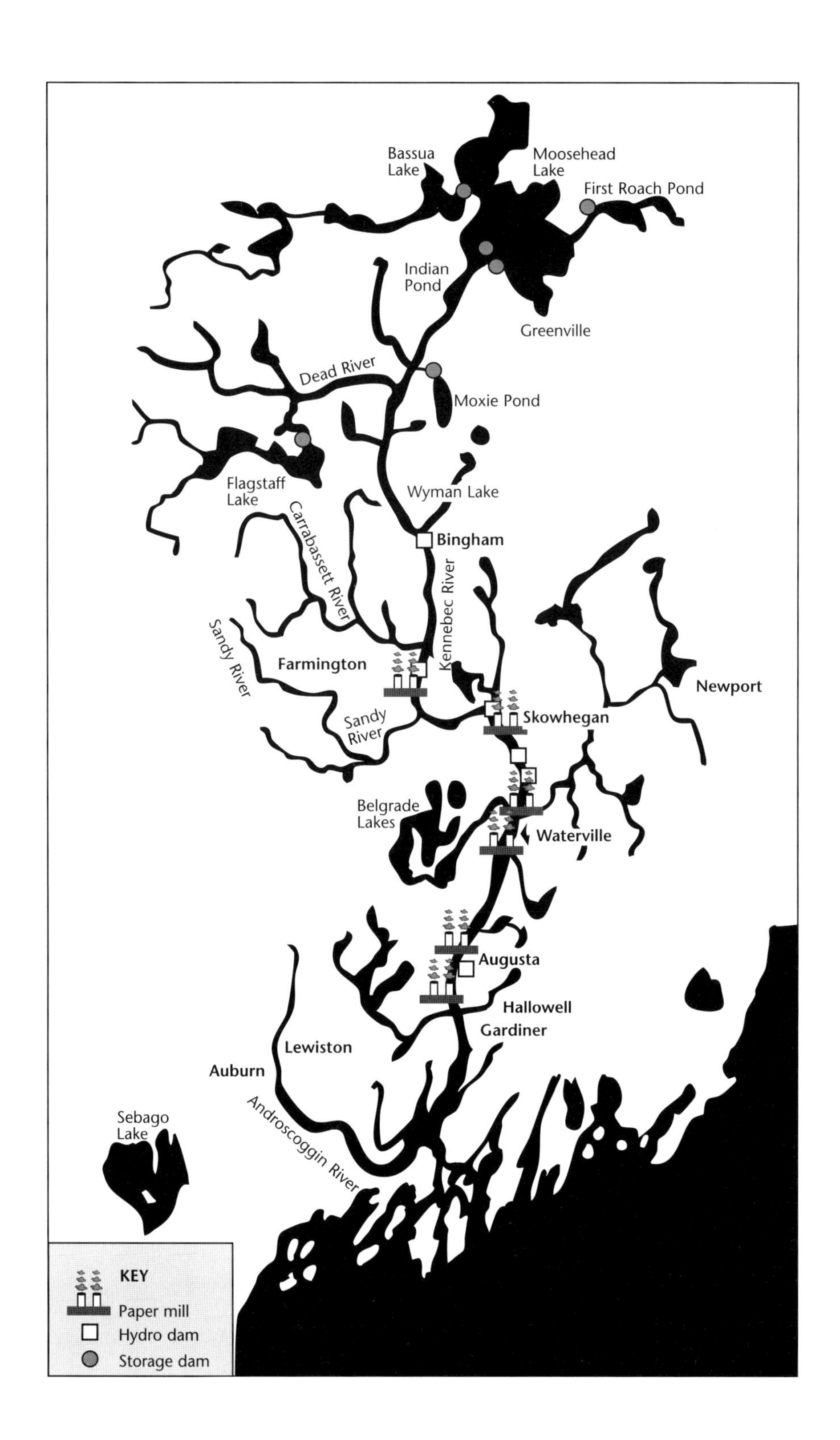

8.16 Dams of the Kennebec River

Dams have altered the water flow regime on every major river entering the Gulf of Maine. On the Kennebec River six major dams are located between Bingham and Hallowell in addition to six storage dams on lakes that are part of the watershed. Although highly important for flood control, these dams reduce the nutrient enrichment function of the rivers as they enter estuarine areas. *Credit: A. Linton, based on information in the Maine Sunday Telegram*

approach to land use within watersheds, and restoring watersheds that have been damaged. In Clemensport, Nova Scotia, for example, the Clean Annapolis River Project (CARP) has started several interdisciplinary projects to restore and preserve the Annapolis watershed. Since 1605, when Samuel de Champlain and a group of European explorers established a settlement in the Annapolis River basin, the watershed has become a major agricultural area and the natural environment has been greatly altered. The Annapolis River Guardians project was designed to train citizen volunteers to monitor water quality within the watershed (fig. 8.17).

The Black River Stream Reforestation project, another CARP project, has documented severe streambank erosion caused by uncontrolled livestock access to the stream. Restoration efforts have included installing fences along banks, planting trees, and limiting livestock access. The landowner of this area signed a stewardship agreement to maintain the area for ten years. CARP was also instrumental in a series of projects in which 700 elementary students within the Annapolis River watershed were questioned on how they use its resources and on their future expectations of the watershed (e.g., potable water, presence of healthy fish). In another project, senior citizens' knowledge of how the region has changed over the past seventy years is being recorded. The purpose of these projects is to help to develop an environmental management strategy for the Annapolis watershed.

In Casco Bay, Maine citizens' groups recently lobbied the Environmental Protection Agency to nominate Casco Bay to EPA's National Estuary Program. With the help of significant federal funding, the program seeks to provide educational information illustrating the link between land uses along the Presumpscot River and water quality in the Casco Bay estuary. Program participants say that this linkage has been clearly established for most townspeople who live adjacent to the bay, but that it will take time for townspeople upriver to understand how their actions have the capacity to affect water quality.

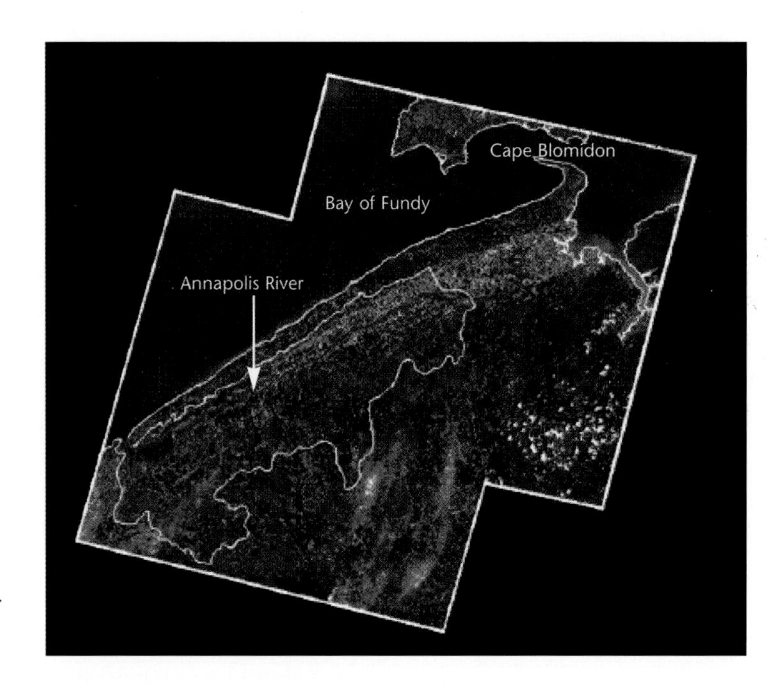

8.17 Annapolis River Watershed

The Clean Annapolis River Project (CARP) has started several interdisciplinary projects to restore and preserve the Annapolis watershed and to present important scientific and technical information about this watershed to the public. One of the projects is designed to train citizen volunteers to monitor water quality within the watershed. *Credit: D. Colville, College of Geographic Science, Nova Scotia*

Ultimately, we are all citizens of one watershed or another, wherever we live on the globe. Watershed consciousness has not seeped very deeply yet into the public mind; and no wonder, when you add up all the watersheds that empty into the Gulf of Maine, a far-flung and diverse set of communities across state, provincial, and international boundaries. In the end, though, all citizens of the region will need to become involved in the challenges and opportunities of maintaining healthy and productive aquatic and marine systems. We cannot afford not to, because the economy and public health of the region will increasingly be determined by how we perceive our common fate.

9

VIEWS OF THE FOREST

TIMBER, HISTORY, AND WILDLANDS OF THE GULF OF MAINE WATERSHED

Philip W. Conkling, Lloyd C. Irland, and Janice Harvey

9.1 Vegetation Index of Northeastern North America

Multiple AVHRR images with a resolution of 1 kilometer were merged to produce this composite image of the distribution of different vegetation types over northeastern North America. This image reveals a number of important meta-landscape features. The dark green areas denote the distribution of boreal conifers such as spruce, fir, and white pine in southern Quebec, the Adirondacks, northern Vermont and New Hampshire, and throughout most of Maine and New Brunswick. Lighter shades of green in the southern part of the Gulf of Maine region represent the distribution of hardwood forests, which intergrade in a broad transition zone across the region from central New York to southern Maine. It is interesting to note the diffusion of hardwood forests in northern Maine, beginning at Moosehead Lake, the region's largest freshwater body (near the center of the image) and extending into the Saint John River valley on both sides of the international boundary. Yellow areas represent Arctic heath and tundra to the north and agricultural and scrub coastal plain vegetation to the south. Both Boston and Saint John, the Gulf of Maine's largest cities, are clearly denoted by large clusters of pixels. *Credit: EROS Data Center, U.S. Geological Survey, and Canadian Centre for Remote Sensing*

The present-day forests of the Gulf of Maine are a product of almost four centuries of human use. No other area of the New World was explored earlier and exploited longer for its forest products than the watershed of the Gulf of Maine. Along with our abundant fish, our forests have been one of the true sources of the region's wealth since the first sail appeared on the horizon. For some ten generations, settlers in the Gulf of Maine region have harnessed the power of the coast's tides and captured the energy in swift rivers to mill old-growth timber. Living off wild game and fish, they pushed deep into the interior to harvest mast pines and "king" spruce. At the turn of this century, they even built mills in the wilderness to produce pulp, paper, and two-by-fours. In these forested watersheds, the use of the trees, fish, water, and wildlife has always been tightly linked, although the finished products and forest practices have changed almost decade by decade.

The region remains one of the most heavily forested in the northern temperate zone, despite its proximity to major population centers at its fringes. Maine is proportionately the most heavily forested state in the United States, with almost 90 percent of its surface covered by trees; New Hampshire is second with 86 percent. New Brunswick's and Nova Scotia's forest cover is also substantial. In much of the remainder of the United States and eastern Canada, forests have been converted to urban and suburban areas or leveled for agriculture. But trees still dominate our landscape from northern Massachusetts to coastal Nova Scotia, as this table suggests:

	Nova Scotia	**New Brunswick**	**Maine**
Total Land (millions ha.)	5.6	7.2	8.0
(millions ac.)	13.8	17.8	19.7
Forest Land (millions ha.)	3.9	6.1	7.1
(millions ac.)	9.6	15.1	17.5
Percent Forest	70	85	89
Annual Acreage Harvested	79,000	250,000	300,000

Figures from Forestry Canada, the USDA Forest Service, and the Maine Forest Service

For centuries, the resources of this forest region have seemed nearly inexhaustible. Only in the last few years have we begun to see the whole forest, and not just its trees. Indeed we are beginning to appreciate that this forest is an ecosystem of interdependent parts—water, wood, and wildlife—in which the functioning of the parts affects the health of the forest as a whole. Today visualizing the forest ecosystem requires integrated and balanced views of its different parts if it is to be preserved for sustainable future uses.

Vegetation Communities in the Watershed

Ecologists call large regional groupings of plants with similar vegetation structure and sharing a similar environment a *biome.* Two major biomes figure in the Gulf of Maine region (fig. 9.1). The eastern deciduous forest extends from the Great Plains east to the Atlantic coast, and from the Gulf of Mexico as far north as central Minnesota and Maine. Here it intersects the coniferous *boreal* forest (or taiga), which extends at higher latitudes across the entire North American continent and sends fingers of spruce and fir down the spines of the Appalachian Mountains. In our region, its characteristic tree species are the white birch, red and white spruce, and balsam fir.

No actual line divides biomes. Instead, there is a broad zone of transition, and the Gulf of Maine watershed lies in one of these zones of transition. In addition to the oaks, maples, hickories, and beech trees, two softwood species—eastern hemlock and the eastern white pine—have their ranges largely within this transition zone. The deciduous forest biome dominates the forest region into western Maine, where it becomes the Acadian forest of northern Maine and the Maritimes. The principal differences are that the Acadian forest has more balsam fir on upland sites, more red spruce compared to white spruce, and more reliable growing season rainfall.

Ecologists subdivide each of these biomes into smaller units called *forest associations,* each named for its major tree species. Each species has evolved adaptations to a given environment in its abilities to deal with such factors as drought, wind, disease, and different intensities of sunlight and shade. Different species sometimes develop means of solving a similar set of environmental problems, and therefore are often found growing in close proximity to each other, but a forest association or type is, biologically speaking, a collection of individuals.

Ecologists customarily describe vegetation communities according to their dominant members, and so in this chapter we will be talking primarily about trees. Nevertheless, each forest type has a rich understory of understory trees, shrubs, flowering plants, mosses, ferns, lichens, and mushrooms that play important roles within their respective communities (fig. 9.2).

9.2 Understory Lichen

Along the eastern coast of the Gulf of Maine, where fog is prevalent, a large number of lichen species grow abundantly underneath spruce forests. These lichens are able to extract moisture directly out of the air, which helps explain why they are much more abundant in nearshore coastal forest than just a few miles inland. *Credit: P. Ralston*

Forest Associations: Past and Present

Throughout much of the Old World, large-scale environmental changes occurred centuries, if not millennia, before they did here in the New World. In the Gulf of Maine region, generations of explorers, travelers, and settlers left detailed records of the original environment they found, often full of valuable ecological information, always full of wonder, and their accounts allow us to look back in ecological time.

One of the early histories of the region, published by Jeremy Belknap in 1792, underscores how carefully colonists considered forest lands. Like many of the settlers of the region, Belknap was a keen observer of nature. He appreciated the connections between forest trees and the soils beneath them in assessing the underlying productivity of the land. His ecological descriptions correlating forest communities with their soils is astonishing to read more than two centuries later:

Pitch pine land is dry and sandy; it will bear corn and rye with plowing; but it is soon worn out.

White pine land is also light and dry, but has a deeper soil and is of course better.

Spruce and hemlock, in the eastern parts of the state, denote a thin cold soil, which, after much labor in the clearing will indeed bear grass without plowing, but the crops are small.

Beech and maple land is generally esteemed the most easy and advantageous for cultivation, as it is a warm, rich, loamy soil, which easily takes grass, corn, and grain without plowing.

What Belknap recognized is that the distribution of tree species in characteristic patterns is a function of the composition of the soils that nourish the trees from beneath. Today ecologists recognize the following forest types in the region.

Oak-Hickory Association

Oaks are the most widespread and the most commercially important hardwood trees in the northern temperate zone. Twenty oak species have commercial value in the United States and Canada, including five found in the Gulf of Maine region: red, white, black, scarlet, and chestnut oaks. Although red oak is found throughout the region, the other species are restricted to the central and southwestern portions, especially eastern Massachusetts. All of these species have been cut for centuries for the region's once flourishing shipbuilding industry. In more recent decades large quantities of red oaks have been cut and exported for high-quality furniture.

Oaks are found on shallow, ledgy, or sandy soils especially in the southwestern portions of the Gulf of Maine region. They are rarely found in pure stands; rather, oaks are typically associated with hickories (shagbark, bitternut, mockernut, and pignut), or with other hardwoods such as maples and beech farther north.

Pine-Oak Association

On the terminal moraine of Cape Cod and the islands of Martha's Vineyard and Nantucket, the characteristic association is pitch pine and scrub oak (fig. 9.3). Two other oaks, the bur oak and post oak, reach the northern edge of their range on Cape Cod's sandy soils. Pitch pine is adapted to the most sterile soils of the region and is outstanding in its ability to recover from fire because it harbors dormant buds at the base of the tree, often protected by basal crooks, that can resprout after all the rest of the tree is killed.

Pure Pines: White Pine, Red Pine, and Jack Pine Communities

Pines are the most valuable conifers wherever they are found in the northern temperate zone because of their long trunks and straight-grained wood, which has always been valued as a premium building material. The masts of Lord Nelson's HMS *Victory* were grown in Maine and shipped across the Atlantic before the Battle of Trafalgar. Whether for masts of the Royal Navy's ships-of-the-line, for clapboards and framing during colonial times, or for doors and window casements in modern homes, pine's reputation for supreme utility is well deserved.

Pure white pine stands are found today throughout the Gulf of Maine region, especially in southern and western portions where it was the typical species to colonize abandoned fields after the region's agricultural boom peaked in the mid-nineteenth century. Natural white pine stands are also found on old glacial outwash plains and along river banks.

White pine is capable of reaching an age of 300 or more years, and old-growth white pine stands are found in a handful of places throughout the region (fig. 9.4), including the Gulf Hagas stand south of Mount Katahdin on the Appalachian Trail. The Bowdoin Pines, a beautiful stand of trees on the campus of Bowdoin College in Brunswick, Maine, is located on an outwash plain created by the glacial meltwaters that rerouted the Androscoggin River. But this is by no means a virgin stand; these were farmers' fields until the mid-1850s.

Northern Hardwood–Hemlock

No other forest type in the Gulf of Maine is more complex and more important to a broad cross section of birds and mammals than the northern hardwoods (fig. 9.5). Although the dominant species in these communities are beech, sugar and red maples, and white and yellow birches, a large number of other hardwoods appear with them. In the northern hardwood stands, large white pines, hemlocks, or red spruce emerge from the canopy. On the forest floor in these communities (which are found on the richest soils of the region), we find some of the most beautiful forest flowers, including the dog-toothed violet, pink and yellow lady slippers, and spring beauty.

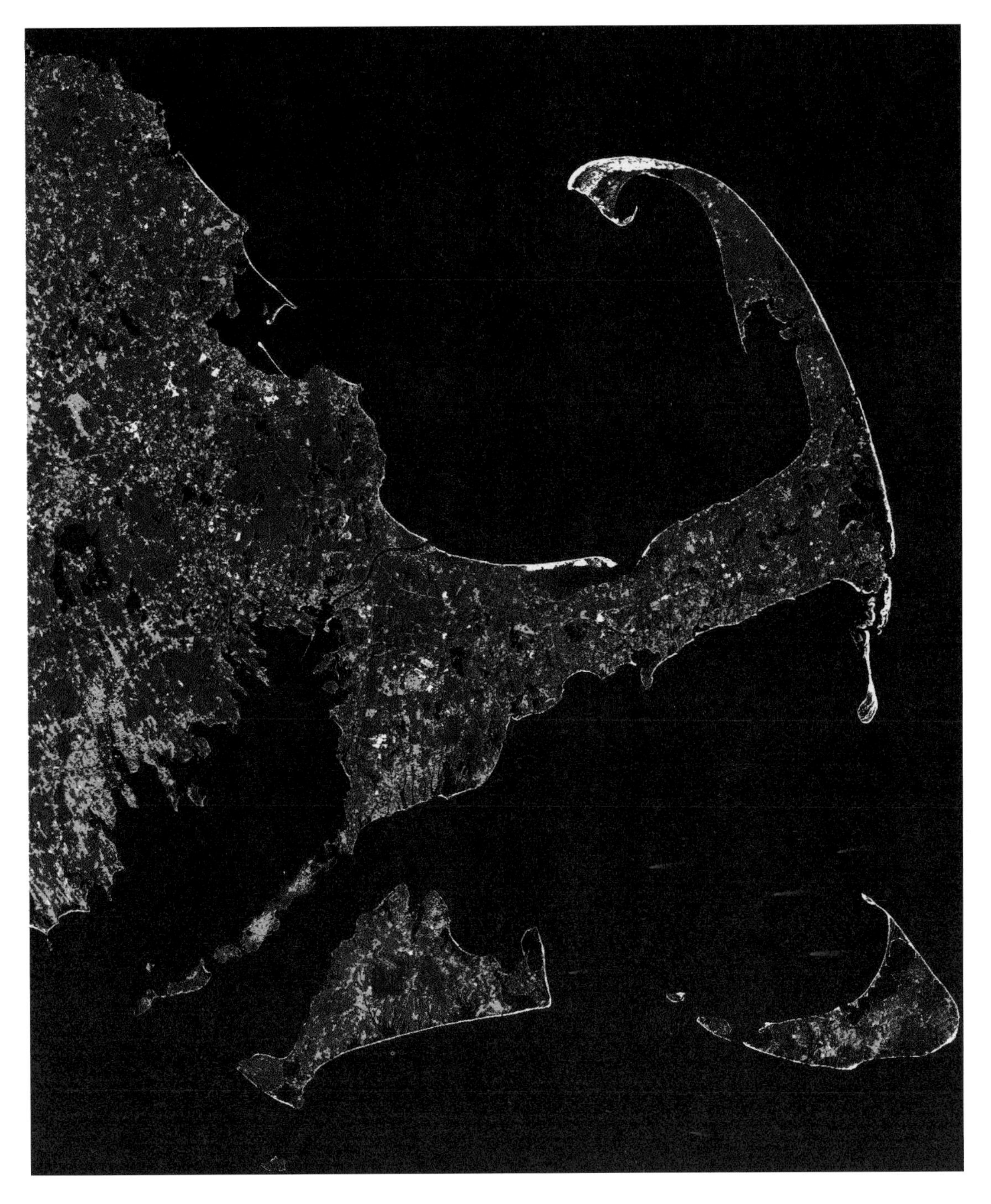

9.3 Cape Cod Pine-Oak Community

This theme-mapped satellite image shows the distribution of white pine and pitch pine (dark green) amid more abundant oaks (orange and brown). Most of the soils in this image are sandy and extremely dry. Relatively few forest species are adapted to this environment. The block of green that appears in the middle of Martha's Vineyard is a pine plantation established in the 1930s. *Credit: P. Conkling, R. Podolsky, Island Institute; GAIA image*

9.4 Broad Arrow White Pine

Ordinances imposed by the king of England on his American colonies reserved all white pine trees greater than two feet in diameter for masts for the Royal Navy's ships-of-the-line. Marked with a broad arrow by a forester's ax, these pines were the most valuable tree in the colonial forests of the region. During a survey of old-growth pine stands conducted by the State of Maine in 1977, the tree pictured here from a protected grove of trees near the Saco River in Hiram, Maine, was identified as a likely surviving broad arrow pine. *Credit: J. Kosinski*

These forests today produce a rich variety of forest products. Sugar "bushes," or stands of sugar maples, which occur both naturally and in plantations, produce one of New England's most famous specialty products. Baseball bats (including the famous Louisville Slugger), sleds, toboggans, and snowshoes are made in Maine and Canadian mills from white ash. The veneer that is produced from yellow birch today for use in kitchen counters and other specialty cabinetry products was prized during the 1940s to make the wings and propellers of England's famous Mosquito bombers. And soft-grained basswood is still the preferred species for woodcarvers, although it is much less important than a century ago, when few wooden ships were launched from Bath or Fredericton without a carved figure under the bowsprit.

Spruce-Fir Association

Three species of spruce—red, white, and black—along with their chief associate, the balsam fir, are the predominant forest cover of vast regions of the Gulf of Maine watershed. All of these species are adapted to areas characterized by intense cold and shallow, acidic soils. Red spruce and balsam fir are adapted to grow in shade, and their saplings can persist in the dark understory shade of older trees for upwards of 60 years. Foresters today still refer to dense spruce-fir stands as "black growth" for their dark interior (fig. 9.6). If you look at old maps and atlases, you find many headlands called Black Point and many islands named Black Island; these old place names have an underlying ecological meaning. The coast of the Gulf of Maine, with its circulating cold currents, creates cooler growing conditions than the coastal plain just to the south. Maine and the Canadian Maritimes are the only places on the Atlantic seaboard where spruce and fir occur at the coastline.

Spruce and fir are premier woods for making paper. Their fibers, when broken down in the pulping process, are longer than those of any hardwood and are an essential ingredi-

9.5 Northern Hardwood Forests

This fall image of northern hardwood forests suggests the diversity of tree species that grow on well-drained sites of rolling topography with deep, rich forest soils. *Credit: P. Ralston*

ent for high-grade papers and for newsprint, which must be strong enough to run through modern high-speed printing presses. In addition, these species also produce prime dimension lumber for framing houses and other parts of the construction industry.

The scourge of spruce and fir is a small endemic insect named the spruce budworm, no bigger than a quarter of inch, which periodically erupts in massive infestations and lays waste to thousands of square miles of softwood timber on which tens of thousands of jobs throughout the region depend. These infestations occur every 60 years or so; the most recent outbreak (1973 to 1983) came during a time when citizens were increasingly concerned about the use of pesticides, including DDT, which had previously been sprayed over large acreages in Maine. Although DDT was banned in 1973, millions of acres of the Gulf of Maine watershed were sprayed with a variety of other pesticides from converted World War II bombers for almost a decade, to allow large forest landowners time to salvage some of this wood on which their mills depended. The outbreak collapsed for reasons that are still mysterious, but it will undoubtedly recur, and the issue of aerial application of pesticides will likely reignite old debates in the region over uses of the forest (fig. 9.7).

9.6 Black-Growth Spruce
(above)

Most of the sunlight that strikes a northern spruce canopy is absorbed by the dense array of needles on these trees, creating a cool, dark understory environment. *Credit: P. Conkling*

9.7 Spraying for Spruce Budworm
(right)

Pesticides sprayed from planes were used to control the spruce budworm infestation in northern Maine, New Brunswick, and Nova Scotia forests for more than a decade in the 1970s and 1980s. This view of a Maine Forest Service spraying operation in 1981 shows two spray planes releasing the chemical insecticide carbaryl, which was sprayed on 1,015,000 acres of Maine forest that year. *Credit: James Sewall Co., courtesy of Maine Forest Service*

Two Centuries of Logging in the North Woods

Two centuries ago the New World's first lumber boom began in earnest, and it was centered in the heart of the Gulf of Maine region. Each year between 1840 and 1880, thousands of lumber schooners sailed up to head tide on the Penobscot and scores of other rivers to load pine and spruce lumber that had been driven downstream from the far reaches of the enormous watersheds upcountry. In Bangor, Maine—lumber capital of the world before the Civil War—the peavey, boom chain, calk boots, and Paul Bunyan were invented. Dozens of sawmills that produced clapboards, boards, and timbers lined the banks of the

Penobscot, just as they did the region's other major lumber waterways, the Merrimack, Piscataqua, Kennebec, Machias, St. Croix, and Saint John.

During this time, logging was fairly benign ecologically. Since only a few large trees were removed, a forest remained behind, often with enough valuable wood to support a heavy cutting only 20 years later. High-quality pine, for example, was topped at the first branches because it was too time-consuming to chop the limbs from the upper end of the tree when another tall pine with a clear trunk was standing nearby. Since horses were used to haul logs to landings, damage to remaining trees was light. Since cutting was done in winter, there was little or no soil disturbance. Since hardwoods would not float, they were left.

Although forests across the entire region produced huge fortunes for the new lumber barons, it was the region's watery highways that made these enterprises possible (fig. 9.8). But river driving was dangerous and difficult work: picking a log jam with a peavey or trying to hold a million board feet of logs in a boom during the spring freshet ended many a hapless life. Those who controlled the flow of rivers across the Gulf of Maine watershed controlled its wealth. The construction of the Telos Canal in 1841 diverted waters from the Saint John River watershed into the Penobscot. Enormous volumes of pine and spruce timber that otherwise would have been driven to Canadian mills arrived in Bangor instead.

During the heyday of logging, contractors remodeled entire drainage systems. They straightened rivers, installed splash dams, and raised upland ponds and lakes, thereby enhancing spring water flows and rendering tiny brooks drivable. This intensive engineering undoubtedly damaged fish spawning habitat in some areas, while it may have increased it in others, where major lakes were raised and enlarged. Dams blocked fish migration, however, and heavy accumulations of bark, "sinkers," and sawdust were left behind in the streams and estuaries when the saws fell silent. The loggers built well. Even today, large log

9.8 River Driving

For over two centuries, the rivers and streams of much of the interior of the Gulf region served as the primary arteries through which hundreds of millions of board feet of timber were driven from distant forest stands at the headwaters of the watershed to distant mills downstream. The year 1978 saw the last industrial river drive in Maine to use the system of dams, sluices, storage ponds, impoundments, and deadwaters that had been constructed throughout the vast Penobscot and Kennebec watersheds. Trout have been restored to some of their former range in the watershed; salmon have not. This archival photograph of one river drive was taken on the West Branch of the Penobscot River. *Credit: courtesy of Bangor Public Library*

and stone cribworks for log booms have survived decades of spring floods and ice jams.

Perhaps the most lasting legacy of the era was the birth of the conservation ethic in the wake of the depredations of the early cuttings. Thoreau was conservation's earliest voice. After a trip to the Maine woods in 1841, he bitterly wrote: "The Anglo-American can indeed cut down and grub up all this waving forest, and make a stump speech, and vote for Buchanan on its ruins, but he cannot converse with the spirit of the trees he fells, he cannot read the poetry and mythology which retires as he advances."

During this period forest fires starting in the dry slash left behind lay waste to tens of thousands of square miles of cutover land in New Hampshire, Maine, and New Brunswick and led to the first forest laws in either country. In New England, the massive cutting and fires that ravaged the flanks of the White Mountains led to the creation of the watershed's first and only national forest after Congress appropriated funds to buy cut-over private lands. The Appalachian Mountain Club was formed in Boston in 1876 to advocate forest protection, and in 1901 the Society for the Protection of New Hampshire Forests was started. But on many of these burned-over lands, red spruce and white pine, the region's most valuable species, were replaced by scrub forests of gray birch, pin cherry, and aspen, which, though pretty enough to look at, would not recover their original composition and former commercial importance for more than a century.

By the early twentieth century, the rising cost of manufacturing paper from rags, together with the greatly increased demand for paper, ushered the Gulf of Maine region into the modern period of pulpwood production, whose corporate empires stretched from New Hampshire across Maine and into New Brunswick and Nova Scotia. Following the formation of International Paper Company on the Kennebec in 1898 and Great Northern Paper Company on the Penobscot in 1900, which moved its headquarters literally into the wilderness to build its own town, Millinocket (fig. 9.9), many of the waterways were revisited by wood crews to cut the smaller spruce and balsam fir that had been bypassed in the earlier harvest of sawlogs.

After 1940, the replacement of the horse by the truck and the crosscut saw by the chainsaw allowed loggers to cut larger areas in a shorter period of time. By the late 1960s, because of concern that freshwater fisheries were severely damaged in many of the streams and rivers where logging waste destroyed natural habitats, river driving was phased out as the means of pulpwood transportation. To replace the river highways, paper companies have embarked on a program to build over 15,000 miles of roads into the woods, opening up areas that were previously too remote from water to be cut profitably. With the new road system have come first rubber-tired skidders and then feller-bunchers, delimbers, and other huge mechanized logging equipment, enabling pulpwood cutters to clear-cut up to 1,000 acres during a single operation (fig. 9.10).

The effect of nearly two centuries of heavy cutting on softwood lands throughout the interior of the Gulf of Maine has been to shift huge acreages out of spruce, fir, and pine growth and into lower-quality hardwoods such as white and gray birch, red maple, and pin cherry. Because pulping processes are capable of utilizing ever greater percentages of these hardwoods, raw material supply is not in question. But those mills that depend on spruce and fir and pine have had to resort to herbicide spraying to control unwanted hardwood regeneration (fig. 9.11).

It is a striking fact that during the logging era almost no voices were raised for retaining any areas of the coast, lakeshores, forests, or mountains in a pristine condition. This was true as well in New Brunswick and Nova Scotia, where extensive forests remain in public ownership even today. Most people of the nineteenth century would have found preservation of virgin forests a strange and even antisocial notion. Progressive people of the first half of the nineteenth century believed the forest was a useful but temporary resource, which would be cleared

9.9 Millinocket and East Millinocket (left)

The pulp and paper mills of Millinocket and East Millinocket were literally carved out of the wilderness along the banks of the West Branch of the Penobscot River shortly after 1900 by the Great Northern Paper Company. Great Northern operated until the late 1980s, when it was acquired first by Georgia Pacific Company and then shortly thereafter by Bowater Inc. *Credit: P. Conkling, T. Ongaro, Island Institute; GAIA image*

9.10 Telos Road Clear-cuts (right)

This satellite image encompasses approximately 100 square miles or roughly 64,000 acres of northern forest northwest of Millinocket. Much of the area was heavily clear-cut by the Great Northern Paper Company during the budworm outbreak of the 1970s and 1980s. Telos Pond, part of the Allagash Wilderness Waterway, is at the top of the image. Sourdnahunk Lake to the right forms the western border of Baxter State Park. Logging roads crisscross the area. The plumes of smoke, visible from space, appear to be from the burning of slash piles. *Credit: P. Conkling, T. Ongaro, Island Institute; GAIA image*

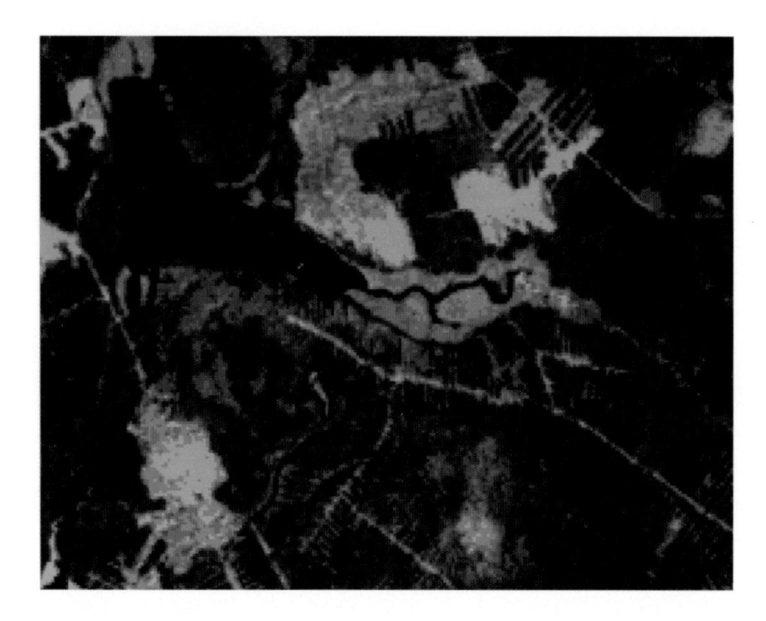

9.11 Strip Cuts Close-up

Great Northern tried numerous configurations of clear-cutting, including narrow strip cuts, which foresters hoped would provide enough shade to reduce hardwood shrub competition and raspberry regeneration. These strip cuts were largely unsuccessful in controlling unwanted vegetation, and the uncut strips were later cut. *Credit: P. Conkling, T. Ongaro, Island Institute; GAIA image*

to make way for farms. As a result of this history, the Gulf of Maine watershed contains only tiny, accidental remnants of forest in a primeval condition, the scarcity of which makes it difficult to gauge the degree to which the biological diversity of the region may have been compromised.

Divergent Views of the Forest

The forest is interwoven with the region's identity and sense of place to a degree not readily appreciated. As havens for wildlife, as the wellspring for the region's lakes and rivers, as places to repair for hunting and fishing, forests affect people in the Gulf of Maine as in few other places in eastern North America. But different people view this vast forest region from radically different points of view; although we may all be looking at the same forest, we see it through different eyes, with radically different implications for future forest management and land use decisions. It may, therefore, be helpful to review some of these distinct views of the forests of the Gulf of Maine.

The Forest as a Suburban Resource

Most of the population here, as in the rest of the western world, does not live in or even near a large expanse of forest. We live in cities, suburbs, and small towns where tree-lined streets provide our most constant connection to the life of the forest. Here, forest birds that can tolerate close association with civilization herald the arrival of spring with sparkling bursts of song. Suburban wildlife such as raccoons, groundhogs, skunks, and porcupines wander through patchy forest lots. People plant trees and shrubs from other regions of the country and world, which escape and colonize the outskirts of town. Neighborhood kids play hide and seek and maintain elaborate trails through seemingly abandoned lots until they are developed. These suburban forests are our most constant reminder of the once and almost forgotten forests of our forebears and provide a pleasant diversity to everyday life (figs. 9.12, 9.13).

9.12 South Portland Suburban Forest

This GAIA student project map of South Portland, a community at the periphery of Maine's largest city, shows the complex mosaic of vegetation habitats and forest fragments characteristic of the suburban areas of the region. In the student's words, "My GAIA project took me roughly 6 months to complete, because I wanted to figure all the land cover in the town I live in, South Portland, Maine. I used the town boundaries to make the shape of this project. To verify the classifications of my 60 light reflectance classes, I went ground truthing with another student on our mountain bikes on the weekends. I started this project in my freshman year in high school, and finished the project in October of my sophomore year." *Credit: A. Marro, grade 10, South Portland High School; teacher, J. Salisbury; courtesy of Gaia Crossroads Project, Bigelow Laboratory of Ocean Sciences*

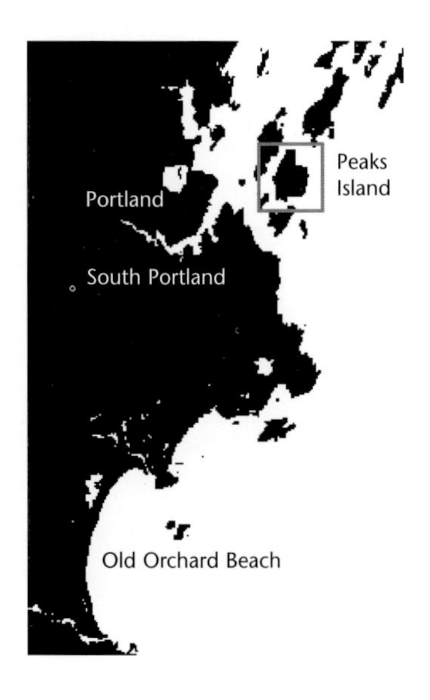

9.13 Peaks Island Forest

Another GAIA student theme map project was conducted on Peaks Island, the most intensively developed island off the coast of Maine. In the words of the student, "To create my Peaks Island map I first worked with other students to make a map of the southern coast of Maine, including wetlands, open space, hardwoods, fields, and other land covers. To verify the maps that I made I went on several ground-truthing trips with my Earth Science teacher. He took us out on mountain bikes on the weekends to do our field work. We each had a color printout of the areas we were working on, and we made notes on that, then put it in the computer." *Credit: C. Smith, grade 9, South Portland High School; teacher: J. Salisbury; courtesy of Gaia Crossroads Project, Bigelow Laboratory for Ocean Sciences*

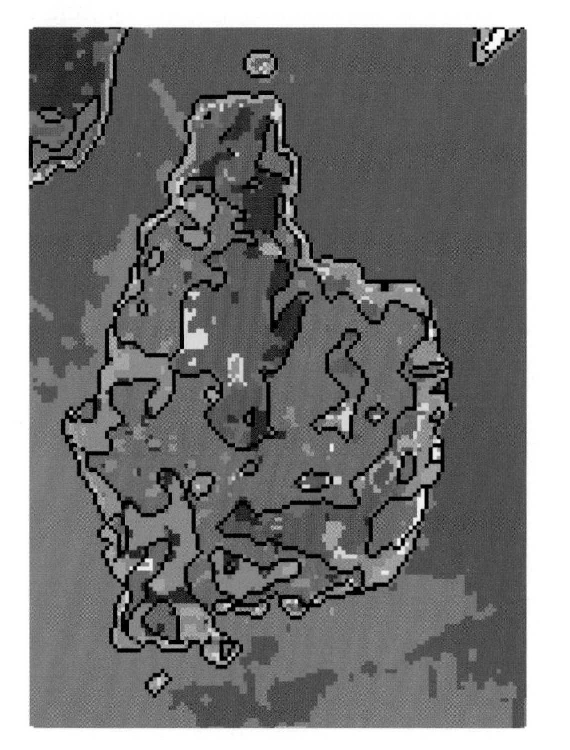

The Forest as a Timber Resource

Compared to other forest values, there are far more abundant and detailed statistics depicting the Gulf of Maine woods as a stockpile of timber and pulpwood, because the forest products industry has been such a major economic asset to the region for such a long time, especially in Maine and New Brunswick. Counts of trees by species, size, volume, county, and condition were and are widely available across the region.

With new technologies, new products, and new markets, spruce and fir trees cut in the northern woods on both sides of the border are now used down to three-inch tops, and tops of small trees are chipped, including leaves and branches. In the last decade, an entirely new industry has developed throughout the region to turn wood chips into energy in biomass boilers. Approximately one-fifth of the electricity sold by Central Maine Power, Maine's largest utility, is generated from wood-fired plants.

As in Maine, the provincially owned lands of New Brunswick and Nova Scotia have also produced rising harvests of timber in recent decades. In the Maritimes, however, tree planting and the culture of plantations is much more extensive than in Maine. In New Brunswick alone, landowner J. D. Irving has planted a total of some 200,000 acres on his private lands.

Gulf of Maine Forests as a Recreational Resource

The interstate highway system in the United States, improved roads in Canada, rising incomes in urban areas, and more roads into the backwoods have fostered a rising tide of recreational hunting, fishing, and canoeing throughout the forests of the Gulf of Maine. Between 1970 and 1980, after rivers were replaced as the principal route logs traveled to market, a vast network of roads was constructed through the northern forests—10,000 miles in Maine alone, according to the state Land Use Regulation Commission. The roads served not only the paper industry but the public in general, putting pressure on natural systems by making the north woods more accessible than ever before. Land use regulators have struggled to keep roads away from key wild lakes and ponds in hopes of keeping their wildness intact and their fisheries healthy. Lots have been carved out of wilderness lakefronts, and properties have been stripped of timber before sale to individual speculators. While the area directly affected has been small, it has had significant impact on the wildness of some remote areas. How best to respond to this part of the recreational industry has been a controversial issue. One statewide group in Maine, the Natural Resources Council, has advocated an outright ban on subdivision and development in Maine's wildlands.

Gulf of Maine Forests as a Wilderness Resource

Forests cover most of the central interior of the Gulf of Maine region and have had a near-legendary quality by virtue of their isolation and sense of wildness. Wildness, of course, is as much a cultural phenomenon as a natural one. The trees of these woods are second and third growth, and some of them are planted. In the Allagash Wilderness Waterway, many visitors arrive by air and the use of motors on canoes is well accepted—hardly the forest primeval! Yet by and large the north woods retain their magical air of wildness, except where large clear-cuts are visible from roads or to passing canoeists (see fig. 8.11). Thoreau spoke of the Maine woods as a wilderness, although even in the 1840s he knew that it was mostly privately owned and slated to be cut over for spruce logs. Still, he understood the cultural importance of large expanses of forest, and urged that a major reserve be created in the Maine woods.

Today, less than 150,000 acres (60,000 hectares) are set aside as federal wilderness on the U.S. side of the Gulf of Maine watershed, mostly in the White Mountain National Forest. Other wilderness areas include portions of Baxter State Park (200,000 acres) and 20,000 acres of the Allagash Wilderness

Waterway. In New Brunswick and Nova Scotia, 110,000 acres (45,000 hectares) are set aside as provincial parks. Canoeing on the Saint John is managed by a cooperative of private landowners instead of by a public agency. But altogether, only a tiny percent of the whole forest is currently set aside. Several cycles of forest use have come and gone, leaving only tag ends of the former virgin forest. A Maine state survey recently found only about 1,500 acres of virgin forest on state land, and only tiny remnants persist on private lands.

In New Brunswick the provincial government proposes to build a road called the Fundy Trail through the last remaining stretch of old-growth timber without public road access (fig. 9.14). This section of the Fundy coastline, adjacent to Fundy National Park, contains stands of old trees, particularly red spruce, some of which exceed 1.5 meters (5 feet) in circumference. A number of conservation and wildlife groups have formed the Fundy Wilderness Coalition to raise awareness of the area, and to oppose the construction of the road.

In place of the road, the Fundy Wilderness Coalition has proposed a network of hiking trails that would bring people to the sensitive coastal bluffs and ravines, allowing them to experience reasonably extensive coastal forest wilderness and undeveloped shoreline. Many of these watersheds have deep-cut ravines, gorges, and valleys, some with sheer cliffs many hundreds of feet high, that empty into the Bay of Fundy.

The Gulf of Maine Forest as a Biological Resource

Hunting pressure and habitat changes have eliminated several of the prominent wildlife species of this forest, including the timber wolf, mountain lion, and caribou. In the more settled portions of the region, however, a significant acreage of today's woodland is growing on former farmlands that were abandoned because they could not compete with farms in other regions. In these areas, species such as the whitetail deer, wild turkey, and beaver are returning to prominence. In Maine, the statewide deer kill has fallen slightly since the 1950s, as habitat conditions have shifted away from those favoring deer. In contrast, from the mid-1970s to the early 1990s the number of active bald eagle nesting sites and young fledged displayed an encouraging uptrend.

In the 1970s, researchers began to detect that rain and snow in the Northeast had become acidified by industrial pollution, even as high-elevation stands of spruce were displaying a synchronous decline along the Appalachian Mountain range from North Carolina to Maine. Scientists have attempted to determine whether this decline is due to atmospheric pollutants, to normal stand aging, to a delayed response to the drought of the early 1960s, to all of these factors together, or to yet another factor as yet unknown. No final consensus has been reached, but it is known that alpine ponds from western Maine to the Adirondacks have become acidified, with accompanying loss of fish and other life. Many scientists continue to suspect atmospheric pollutants, but a convincing mechanism or set of mechanisms has yet to be generally accepted (fig. 9.15).

Forests of the Future: Wilderness, Wildness, and Working Landscapes

These five views of the forests of the Gulf of Maine do not fully exhaust the possibilities, but each represents a critical and conflicting resource we expect the forest to provide. The problem is not how to choose which should have absolute primacy, but to recognize that the value in each of these perspectives is likely to increase in the future.

Not only are the issues affecting the forest changing, but the political and institutional forces dealing with forest policies are changing. As an example, a Green candidate ran for governor of Maine in 1994 on a platform advocating major forest policy reform. Another candidate advocates extending the Appalachian Trail north from Mt. Katahdin to the mountains of the Gaspé Peninsula. The petition to list the Atlantic salmon as

9.14 Fundy Wilderness Area

Old-growth spruce stands (in dark green) in this satellite image comprise over half of Fundy National Park in the upper Bay of Fundy region. Elsewhere in this satellite scene, large blocks of contiguous old-growth forest are comparatively rare. Hardwood lands are shown in orange, and wetlands (both freshwater and marine) in purple. *Credit: T. Ongaro, Island Institute; GAIA image*

an endangered species raises the specter of a significant new federal presence on the American side of the Gulf. Once the province of a few interested legislators, industry lobbyists, and executives, forest policy has been opened up to new political forces and groups. Similar policy debates are under way in New Brunswick, Quebec, and Nova Scotia.

History shows that policy issues affecting our forests are never finally resolved. Rather, when conditions demand it, the political process generates an uneasy compromise. Each particular compromise may persist only briefly, or it may endure for generations. We are now challenged to rebuild many policies, to meet the needs of new "ecosystem management" or "landscape management" paradigms.

Perhaps the most significant challenge for the future is not whether a sufficient quantity of forest trees and forested acres can be maintained in the region, but how the quality of these critical regional resources will be maintained. Forest scientists Robert Seymour and Malcom Hunter, Jr., of the University of Maine have advocated a "triad" concept to assist in managing this issue. They argue that there is a role in the forest landscape for more areas of true wilderness. They propose a large area in which carefully tailored timber management coexists with wildlife and recreational management, and other areas in which forests are managed to a high degree of intensity for fiber production.

The areas involved would not necessarily be rigid blocks, with these different management objectives completely segregated, but would intermingle across the landscape. Of course, there is abundant debate on whether this is the right strategy, and there are difficulties in applying it in practice, but it is encouraging that the debate is occurring.

Although there is no spotted owl in our forests, the demands on private forest landowners to satisfy broader social goals, whether for recreational access to remote wilderness ponds, for the protection of biodiversity in these northern forests, or for enhanced water quality, are unlikely to diminish in the future. The recent connections between the use of chlorine to bleach paper and the appearance of significant levels of dioxin in Maine lobsters only serves to underscore the absolute necessity to see the region as a tightly linked ecosystem of interdependent parts. The forest provides the specific areas needed by a variety of species for food, shelter, mating grounds, and cover from predators. To sustain its ecological health demands not only an intimate knowledge of its parts but a view of the interdependent whole.

9.15 Acid Rain Damage in the White Mountains

Satellite imagery can detect plant tissue damage (shown here in pink) due to acidic rain and fog deposition. Acid fog is actually more damaging than acid rain because the smaller droplets in fog can penetrate into leaf tissue more readily than can raindrops and fogs are often more persistent than rains. Alpine zones just below timberline (shown here in white) are also in the direct path of what researchers have termed the "atmospheric sewer," a polluted band of air between 3,000 and 6,000 feet in altitude that carries sulfur emissions from the industrial Midwest across our region. *Credit: B. Rock, G. Lauton, University of New Hampshire*

10.1 Living Tapestry of Form

Credit: P. Ralston

10

BIODIVERSITY

ANIMAL, PLANT, AND LANDSCAPE DIVERSITY IN THE GULF OF MAINE

John J. H. Albright and Richard Podolsky

What is the view of the landscape of a yellow-rumped warbler working its way up the Atlantic seaboard into the Gulf of Maine watershed? What does it see when it looks down? What does a blue-spotted salamander, warming in a soft April rain, encounter when it ventures out to seek its natal vernal pool? Or what would the first mourning cloak butterfly sense, emerging from its hibernaculum beneath a slab of bark in April?

And what of the plants? Could we even imagine the infinite spectrum of changing shades of green without witnessing it each season? What is the connection between this thin verdant mantle and bugs, birds, and bears, which together comprise the wonderfully diverse and living tapestry of form, color, and motion of the Gulf of Maine (fig. 10.1)?

Our perception of the landscape depends on our particular vantage point. At different scales, and with different life strategies, all plants and animals, including humans, are inextricably linked in the mutual struggle for survival. And even though in the great scheme of things species come and go, we depend on the interplay of this rich diversity of animals and plants to continue to bind our natural world together for millennia.

But will it? What is happening to the fecund and evidently fragile fabric of biological diversity of which we are part? Growing evidence tells us that diversity in natural systems is critical to our survival. But what exactly is this diversity? How do we measure it? How shall we conserve it?

Answering these questions, and implementing strategies to protect native biodiversity, are central themes of the new conservation science. This new science marries the elegance of classical natural history study and observation with the space age technology of computer-generated satellite images. From the dry heathlands and sandplains of Cape Cod to the saturated peatlands of northern Maine and Nova Scotia, the challenge is to understand the living diversity of natural systems, and to use our new perspective to regain an appropriate balance in what we take from, and give back to, the land.

10.2 Sensitive Marine Zones of Eastern Maine

Mudflats where nutrients collect and settle out in fine-grained sediments (shown here in yellow) are distributed at the heads of bays east of Mount Desert Island. *Credit: R. Podolsky, S. Meyer, P. Conkling, Island Institute; GAIA image*

Catastrophic natural disasters, over which we may have no control, are normal ecological events with which animals and plants and landscapes have evolved. Remember, after all, that not more than 12,000 years ago the Gulf of Maine watershed was under a mile of ice, so the landscape and species diversity we observe today are relatively fresh and new. Even extinction is a natural event, and has been an important influence in shaping the biological diversity we enjoy today.

What, then, is so different that causes alarm? The answer, simply put, is *pace*. We have increased the rate and, as important, the frequency of landscape change beyond what many native plants and animals can accommodate. This includes even those land uses that we presume mimic natural processes. Less obviously, but perhaps more insidiously, we are undermining the ecological integrity of natural ecosystems through conversion, fragmentation, and simplification. As a result, we are limiting the future survival options of plants and animals by limiting the future availability and quality of suitable landscapes.

Fortunately, at the same time we are developing new technologies that help us observe the natural world. Our growing awareness of native diversity, through careful study and observation at the levels of cell, organism, and community, is now augmented by an unprecedented ability to observe and analyze entire landscapes on a computer screen (fig. 10.2). These digital mapping capabilities can help us chart the ever-changing landscape and determine how to rectify the insensitivities of the recent past.

The Elements of Diversity

Biological diversity (*biodiversity*) is usually defined at three levels of biological organization: genes, species, and ecosystems. In short, biodiversity encompasses all the living elements in the landscape. Genes, of course, are the organic template for living organisms: butterflies, soil microbes, plants, and animals look and behave the way they do because of their particular combination of genes. What Gregor Mendel discovered with his peas, and what we sometimes forget, is that the ability of the organism to live and adapt to change is a direct function of the diversity of genetic material contained within the nucleus of each cell.

We "see" genes expressed in the dazzling variety of shapes, colors, and sounds of individual species of plants and animals in nature. Because it is easier to comprehend diversity at the species level, it is understandable that our desire to conserve diversity has traditionally focused on species. All wild species—plants, invertebrates, and vertebrates—provide the backdrop of diversity for our own more urban lives. It is appropriate, then, that the traditional definition of wildlife, which grew out of early concern for certain species that were hunted or persecuted to extinction, is now making way for a more inclusive definition that includes all wildlife. Just as oxygen, carbon, nitrogen, and all the other natural elements build our physical world, so do the elements of natural diversity—plants, invertebrates, and vertebrates—build our biological world. Species of plants and animals, in turn, congregate in associations, or natural communities, and natural communities together constitute ecosystems. The total diversity of genes, species, and ecosystems yields a diverse landscape full of the potential to change, and the ability to adapt to change.

It is logical that conservation scientists and ecologists have thus concluded that we cannot be satisfied to save individual species. Species interact with their physical and biological environment (fig. 10.3). These ecological processes, or how species react to the vagaries of climate or random fire, for example, shape the structure and composition of natural communities, and are critical to the survival of those species and the functioning of natural systems. As a result, the concept of conserving ecological processes is becoming prominent in landscape conservation planning. Current conservation models consider not only the component species but also the dynamic interactions occurring within and between natural systems.

10.3 Fireweed

The wind-distributed silky seeds of fireweed carry it around the region, where it flowers in splendid profusion, particularly when it lands on burned ground. *Credit: P. Ralston*

The Individual Struggle for Survival

The survival of any organism depends on its being able to obtain food, water, shelter, and space. The traditional definition of habitat includes these four important requirements. Natural communities and ecosystems constitute the available habitat not only for animals but for plants as well. Many species are considered habitat specialists, that is, they have specific requirements that can only be met by one or only a few specific habitat—or natural community—types. Probably many more species are habitat generalists, whose specific requirements can be met by a wide range of natural community types. How specific natural community or habitat types occur in the landscape, then, will influence the composition and abundance of species populations.

It happens that the *ecotone*, the transition zone between two or more habitats or natural community types, frequently shows greater species diversity than the adjacent habitat types. The reasons are several. Perhaps the simplest is that many mobile animal species occupy more than one habitat type during the daily cycle, and can be observed as they pass through the ecotone. Plant species richness, and hence the horizontal and vertical structural diversity of the habitat, tends to be higher in an ecotone, and thus provides more available niches or cover.

The most obvious ecotones are riparian zones along streams and rivers. Equally important are the myriad ecotones between the many patches of natural communities that comprise the landscape mosaic. Satellite images can help biologists locate ecotones and edges, especially the nonriparian ecotones, that may not be as obvious on the ground (fig. 10.4).

Each individual animal and plant, to be sure, faces a seasonal gauntlet that threatens its survival: competition for food, space, and shelter; predators, parasites, and disease. Each also faces both regular and random change that can have profound effects on individuals and on populations. Human land use practices can also alter the size and shape of habitat patches and change the availability and configuration of ecotones. Land use

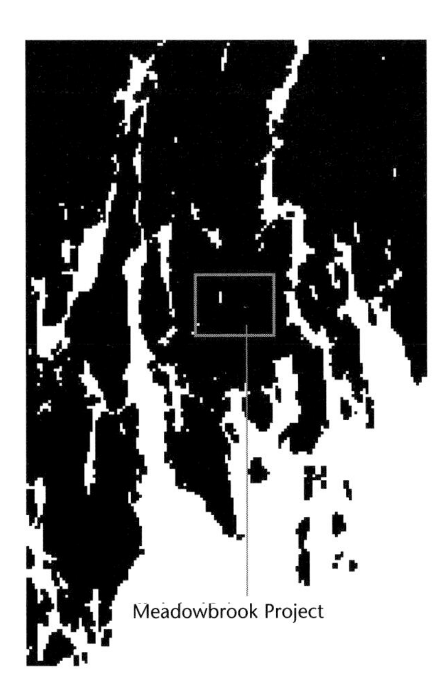

10.4 Boothbay Meadowbrook Area

This GAIA student project nicely illustrates the complexity, diversity, and texture of the landscape in the area around this coastal town midway along the rim of the Gulf of Maine. According to the class, "We wanted to focus on a natural area near our school that we could get to easily. We were interested in finding out what was there, and how it looked on a satellite image. We went out to our study area a couple of times, and did a lot of working together at our computers. We wanted to do acreage tallies to see how much of everything there was and to see what land covers were most dominant." *Credit: students of grade 9, Boothbay Region High School; teacher: J. Clay; courtesy of Gaia Crossroads Project, Bigelow Laboratory for Ocean Sciences*

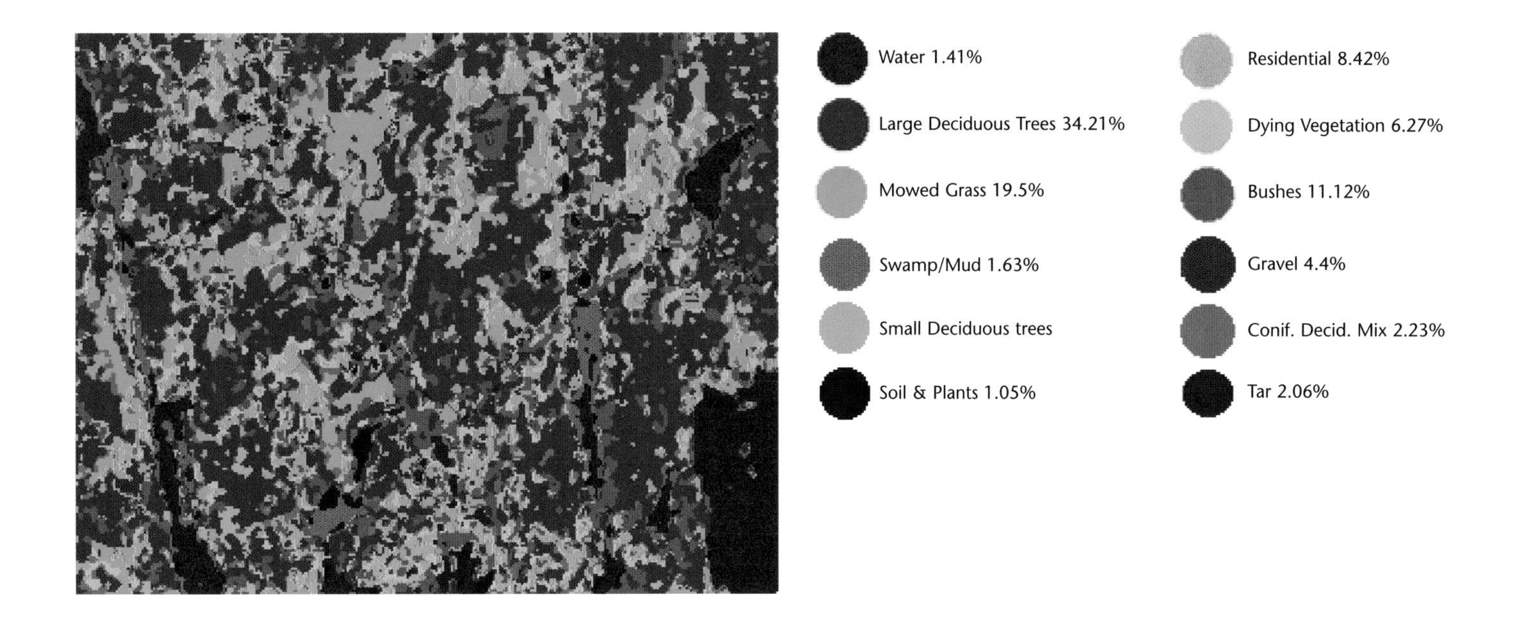

can be critically important to species' survival, both for the specialists that occupy the interior of the patches and for those that occupy or use the ecotones.

Each species has a specific strategy for surviving the regular change of seasons. Bears, skunks, and mourning cloaks hibernate; herbaceous plants and deciduous trees and shrubs shed their leaves or "retreat" altogether underground (along with salamanders, turtles, frogs, snakes, and trillions of seeds, spores, microbes, worms, eggs, and beetle pupae). Others, like warblers, tanagers, monarchs, and some bats, simply leave with the intent of returning to suitable breeding ground the following spring. Deer, on the other hand, move across the landscape from summer to winter grounds, seeking different habitats for food and shelter (fig. 10.5).

Chance or *stochastic* events, like windstorms, floods, and fire, alter or eliminate small or large areas; disease and infestations can slowly or quickly change the shape and functioning of natural communities. At a micro scale, say from a salamander's perspective, the result can be dramatic, virtually eliminating entire suites of plants and animals and allowing a different set of species to occupy the new habitat. At a macro scale, for example that of a wide-ranging coyote, an eagle, or a 10,000-acre forest, site-specific changes can appear insignificant, either because a species is sufficiently mobile that it can move away from the disturbances, or because a large ecosystem can absorb the disturbance. The ability of an ecosystem and its component species to adapt to or wait out site and landscape disturbances could be called ecological resiliency. This resiliency must be preserved along with individual species if we are to enjoy a viable and diverse natural world.

The Geometry of a Changing Landscape

Scale and perspective are important. Small or relatively immobile animals or plants, such as salamanders or orchids, live their lives at a different spatial and often temporal scale than larger or more mobile species, such as birds or bobcats. Our own human perspective is equally important: we must see at all scales to understand the conservation requirements of the full spectrum of biodiversity.

No two species overlap completely in their requirements, which helps guarantee coexistence. Excavating a rotting log in the forest would reveal a diversity of insects all seemingly vying for the same resources. But upon closer examination each species will be seen as more or less specializing on a unique subset of resources, referred to by ecologists as an *ecological niche.* We know that no two species can occupy the same ecological niche at precisely the same time, because eventually one will outcompete the other. The rich, heterogeneous landscape of the Gulf of Maine, at both micro and macro scales, assures an abundance of niches, and hence a diverse and rich landscape of plants and animals (fig. 10.6).

Consider a forested landscape that has been shaped only by natural processes. Such a forest, which provides a myriad of ecological niches in the ground litter, on trunks, on leaves, in branches, under rocks, and on rotting logs, supports a diversity of life that has evolved to live in old forests (fig. 10.7). But a windstorm can fell an ancient tree in an instant, and suddenly sunlight streams to the ground where it has not been for three hundred years. Almost miraculously, an entirely new and perhaps even more species-rich community of plants and animals springs forth, choking the opening and straining to the sun.

Toppling one or several trees within a forest represents a tremendous change for that particular forest patch. But for the forest, it is not unusual. Trees fall randomly from old age and disease, or in larger swaths as a result of fire or wind. Over the entire forest, and over time, these patches continually move; as one patch fills in with growing trees, another opens up nearby. The result is a phenomenon we could call a shifting steady-state mosaic. The kaleidoscope of different-aged patches constantly shifts in response to site-specific disturbances, but the result remains a steady state at the ecosystem level: a vibrant, living, and diverse forest.

Summer Deer Habitat

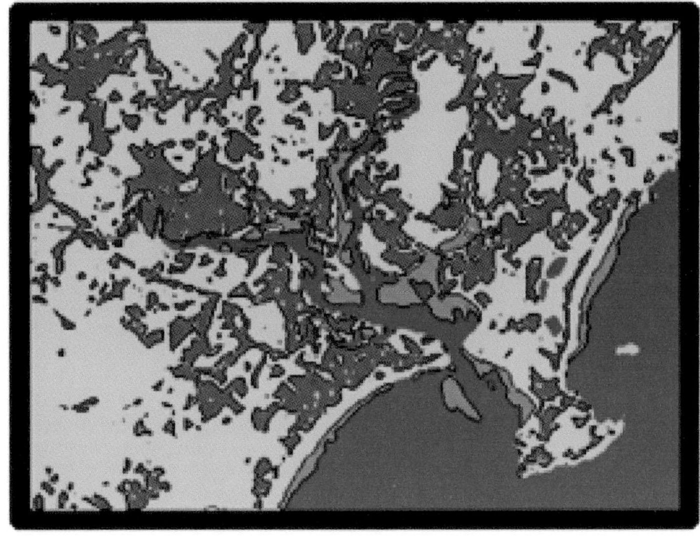

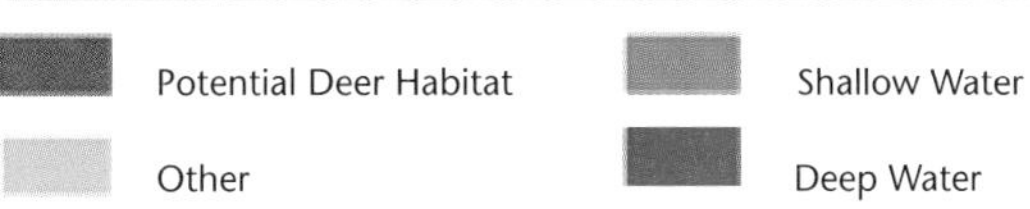

Winter Deer Habitat

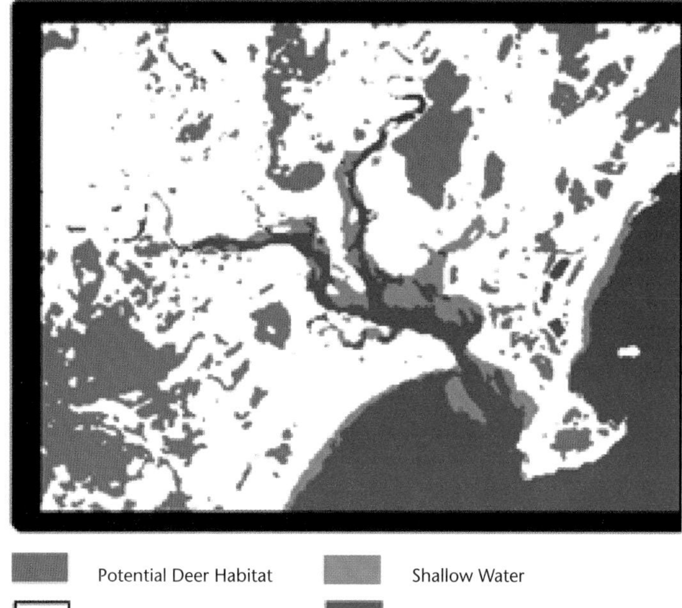

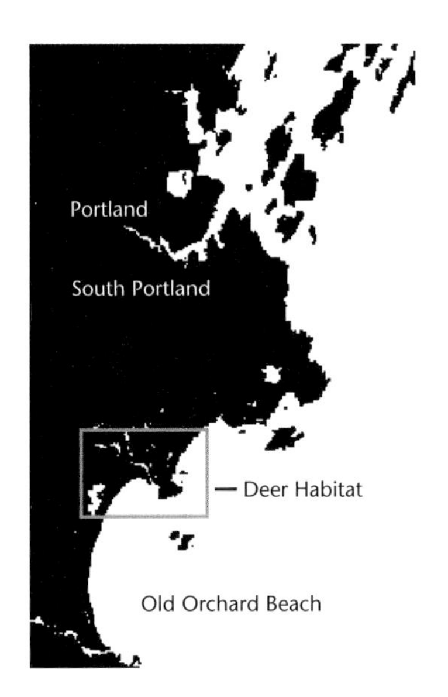

10.5 Summer and Winter Habitat of Deer

Throughout the spring and summer, white-tailed deer forage on hardwood sprouts and shrubs, and in the fall they add weight by feeding on acorns, beech nuts, and hardwood sprouts. In the winter, deer will limit their movement to a "deer yard." This yarding up typically takes place at the margin of a frozen-over forested wetland, especially a cedar swamp, which provides critical winter feed. The yard is usually as far from roads and human disturbance as a given area will allow. *Credit: E. Panenka, C. Smith, and A. Marro, grade 10, South Portland High School; teacher: J. Salisbury; courtesy of Gaia Crossroads Project, Bigelow Laboratory for Ocean Sciences*

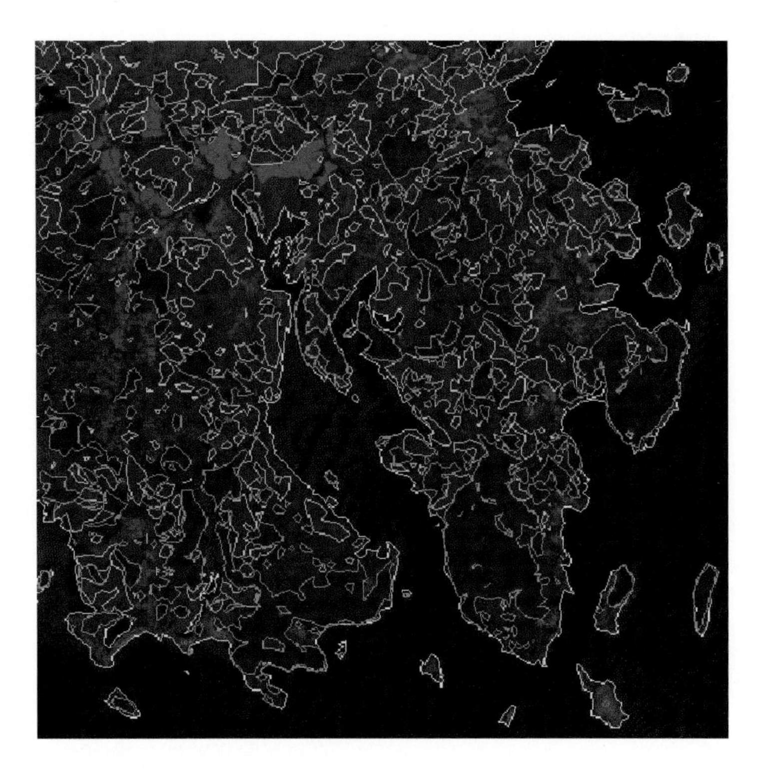

10.6 Landscape Requirements of Deer and Moose

This satellite image shows uplands used by white-tailed deer (outlined in yellow) and lowlands used by moose (bordered in purple). Anywhere in this image where purple overlaps or borders on the yellow is an ecotone between these two habitats.

Large mammals such as moose and white-tailed deer have overlapping though different habitat requirements. Deer are browsers and thus seek out young woody growth, adding forest seeds to their diet in the fall. As a result, their preferred habitat for most of the year is mixed-wood or upland forests. A landscape optimized for deer has an abundance of second-growth forests interspersed with meadows and tracts of dense softwood stands for their winter yards. In contrast, moose specialize during the growing season almost exclusively on submerged aquatic vegetation that grows in the shallows of lakes or slow streams, and hence they frequent bottomlands and wetlands. An optimal landscape for moose is dotted with shallow lakes that are woven together by a rich network of slow-moving streams. However, both deer and moose move through a wide spectrum of landscapes throughout the year, in contrast to small mammals which mostly stay within a single habitat. *Credit: R. Podolsky, Island Institute; GAIA image*

What is the effect on animals and plants? Site-specific change—random or not—may result in the extirpation of a species or set of species from an area. Species that are inherently rare are especially susceptible to such events, which emphasizes the need to identify their whereabouts so that at least we humans do not inadvertently cause their loss. Usually, however, plants and animals are able to wait out or move from the disturbance. Many plant species can lie dormant for many years as corms, bulbs, or seeds. Other species may simply move the 10, 20, or 100 feet to more suitable habitat adjacent to the disturbance. Obviously, the chance events of patches moving through a forest are of less concern to large species (deer, moose, bear) and mobile species (birds). The size of the disturbance is also important. Small patch disturbances may cause a net increase in the overall diversity of the forest or landscape within which the

10.7 Old-Growth Yellow Birch

Undisturbed natural assemblages are rare throughout the region after centuries of human use. This yellow birch tree found in a birch grove on Allen Island in Muscongus Bay, Maine, exceeds five feet in diameter at breast height. *Credit: P. Ralston*

patches move. Very large, catastrophic disturbances can cause local extinctions, and can change the course of ecosystem and landscape development.

Scale is important in another dimension: time. Paleoecologists have pieced together much of the vegetation history of the Gulf of Maine by studying pollen grains buried in sediments of lakes and ponds and in peat. In response to climatic changes during the last 12,000 years, individual species and populations have shifted across the landscape, with resulting shifts in the appearance and geographical distribution of communities and ecosystems. Many of the plant associations we see today are relatively new to the Gulf of Maine, having come together only within the last several thousand years. Changes in climate, and the cumulative effects of long-term catastrophic natural disturbances, will undoubtedly have a profound effect on the species and communities we see today.

Biodiversity Losses

The great auk, a resident and migrant among the islands of the Gulf of Maine, was the first species to go extinct in the New World. This magnificent seabird stood over two feet tall, and was the Northern Hemisphere's equivalent of a flightless penguin (fig. 10.8). Great auks congregated on their nesting grounds in such extraordinary numbers that they seemed an inexhaustible resource to the fishermen and sailors who collected them for food and bait beginning in the late sixteenth century.

Two centuries later the sea mink, endemic only to the coast of Maine and nearby New Brunswick and valued for its fur, suffered a similar fate. A larger relative of the mink still common throughout the northern woods, the sea mink had been trapped for millennia by indigenous people of the Gulf, as evidenced by the number of its bones found in coastal archaeological sites. But the increasing sophistication of trapping methods that the European settlers introduced spelled its doom; the last sea mink was trapped from the coast of Maine sometime in the 1860s.

10.8 Great Auk

The great auk, a flightless seabird of northwest Atlantic waters, was the first species to go extinct in the New World, a victim of overexploitation for food and bait. Archaeological deposits show that the great auk migrated throughout the Gulf in the spring and fall, where it was regularly hunted by indigenous people. Studies of the bone fragments suggest that the great auk may have nested on a few of the outermost islands of Maine. *Credit: courtesy of J. Morehouse*

Today we can only speculate on the loss of biodiversity that often accompanies the destruction of old-growth forests. Virtually all of the forest that stood in the Gulf of Maine region when Europeans arrived in the 1600s is now gone. Even the "wilderness" areas that remain have been logged in the past and amount to fragments too small to provide much true interior forest free from edge effects.

Timber wolves and caribou disappeared from this region by the twentieth century, not because of overhunting but because their habitat had become too fragmented. Fragmentation of habitats and loss of diversity have implications for the stability of all ecosystems, both on land and in the ocean.

Humans as an Ecological Force

If change is so much a part of natural systems, do we need to be concerned with the changes we see in the natural vegetation that covers the land? Do we need to be concerned with the declines of neotropical migrant songbirds and blue spotted salamanders and other amphibians in the Gulf of Maine?

The answer, of course, is yes, and the problem is once again one of scale. With photographic and computer-enhanced images from space, we can for the first time begin to "see" with the proper landscape perspective and to comprehend the real and anticipated impacts of the evidently inexorable spread of urbanization and large-scale land conversion. We humans have succeeded in causing landscape-scale shifts in the structure and functioning of ecosystems. Perhaps more importantly, we have succeeded in doing this more quickly and more frequently than would likely occur in natural systems. As a result, species and communities have less time to adapt and survive.

Loss of biological diversity results from conversion, fragmentation, and simplification of natural systems. Conversion of fields and forests to housing subdivisions has an obvious effect of eliminating natural communities and ecosystems. Converting one ecosystem type to another type—an old forest to a young one; a sedge meadow to a deep marsh—can reduce biological diversity, or can shift the balance in favor of one set of species over another. For example, managing a forest stand with frequent harvests can maintain an early successional forest dominated by aspen and poplar. This approach can be good for deer and grouse, but in the process we eliminate the older beech trees that provide mast for squirrels, bear, and turkey. Converting a streamside meadow into a shallow emergent marsh may be good for waterfowl, but could have profound effects on the much more diverse flora and invertebrate fauna.

Fragmentation of large expanses of eastern forests has resulted in a general decline of neotropical migrant songbirds—birds that breed in temperate and subboreal North America and winter in the tropics. Birds such as warblers, tanagers, thrashers, and even robins exhibited downward trends in numbers from 1966 to 1988, based on analysis of data from 11 breeding bird survey routes from southern New Hampshire to central Maine. Many of these birds are interior-nesting species, like tanagers and vireos, that evidently require an unbroken forest canopy. Fragmenting forest habitats, especially in proximity to urban areas, provides stepping stones of early successional habitats that enable competing bird species or predators to invade the forest interior and reduce the productivity of the nesting migrants. Fragmentation doesn't have to be in large blocks. Nest predators—mammalian, reptilian, and avian—can also use roadways and house lots to invade interior forests (fig. 10.9).

Fragmentation, however, is a tricky issue, and whether it is good or bad depends on both spatial and temporal scales. The patches caused by natural disturbances are a form of fragmentation that occurs at a scale that is accommodated by the forest. Larger forest management activities like clear-cuts and strip cuts are purported to mimic natural processes. This may or may not be true, depending on how often the cutting occurs, and how much of the forest is cut. The line separating patch dynamics and fragmentation can be difficult to discern.

10. 9 Forest Fragmentation around Portland, Maine

Intensively developed and cleared lands (in red and yellow) around Portland indicate the dimensions of human alteration of the landscape. Relatively undisturbed patches of forest lands (in green) are rare in this satellite scene except on the islands of Casco Bay, which may serve as refuges for woodland species, especially migrating songbirds. *Credit: R. Podolsky, P. Conkling, Island Institute; GAIA image*

Simplification of natural systems refers not only to species richness and diversity but also to the physical structure of natural communities and ecosystems. The most famous examples are agricultural systems that yield large areas of single-species "ecosystems." In these systems, simplification occurs at all levels of organization: ecosystem, species, and genetic. Early catastrophes from large-scale crop failures, such as Ireland's potato famine in the 1840s and the corn blight disaster in the United States 100 years later, however, pointed out the danger of such simplification, especially the dangers inherent in a lack of genetic diversity. With a less diverse genetic make-up comes a lessened ability to adapt and a greater susceptibility to pests or disease. The lesson is the same whether for corn, cranberries, or Christmas trees—diversity at all levels is important to the health of natural systems.

Cataloging Biological Diversity

These concerns are a major impetus for conservation action. Obviously, however, it is important to know what it is we want to conserve. Traditional species-specific or site-specific conservation action depended on finding a particular stand of rare wildflowers or the eagle nest or tern island. Amateur botanists or birders, or other naturalists with the "nose"—the apparently innate and elusive ability to find unique species and communities when others cannot—were important for early conservation efforts. But our information needs have expanded with the scope of our conservation goals.

Consider that three to four hundred species of birds, mammals, reptiles, and amphibians breed in or occupy habitats within the Gulf of Maine watershed at some time during the annual cycle. For many of these, distribution, habitat requirements, and conservation needs are fairly well known. But consider that three to four *thousand* species of vascular plants occupy the same area. Add to this the uncounted terrestrial invertebrate species and the task quickly gets out of reach of even the nosiest naturalist.

In tropical countries, where both biodiversity and the rate at which it is being lost are mind-boggling, the conservation strategy is simply to conserve the largest land area possible and then get to the task of identifying what has been conserved. This approach is an extreme example of the "coarse filter" theory of biodiversity conservation. This theory, which is also generally accepted as an appropriate strategy in temperate regions, says that if we can protect representative examples of all ecosystem types, then we will have protected all the native biological diversity occupying those ecosystems.

We are still faced, however, with limited resources of time and money. Largely in response to the need for ecosystem-level information that includes an understanding of the resident species within a particular system (to ensure that we aren't missing rare species or especially diverse local communities), conservation scientists across the country are simultaneously developing a landscape analysis approach to identifying priority areas for conservation. This approach combines the basics of skilled natural history observation and high-technology information management systems. As we begin to understand individual species, or communities of plants and animals—what they need for soils, landscape position, and moisture requirements—we can begin to use existing information about the landscape to predict where target species or natural communities might occur. The success of this approach can be dramatic.

For example, the small whorled pogonia *(Isotria medeoloides)* is one of the rarest orchids in the eastern United States, ranging from North Carolina well into the Gulf of Maine watershed (fig. 10.10). Listed as endangered by the U.S. Fish and Wildlife Service, it was the focus of many thousands of hours of searching in the field for additional stands or populations—with relatively little success. Then, when a small group of ecologists began pooling site data from many states, a picture began emerging: *Isotria* seemed to occur on a particular soil type, in association with intermittent stream channels, on moderate to steep south-facing slopes. The new search image was tested by

10.10 Small Whorled Pogonia
(above)

This small, rare orchid is restricted to areas of calcium-rich soils. Such soils can be mapped, easing the search for important areas of biodiversity. *Credit: G. Cranna, Maine Critical Areas Program*

10.11 Significant Wildlife Habitat Areas
(right)

The Maine Department of Inland Fish and Wildlife has compiled field inventories of breeding seabirds, seal haulout areas, shorebird migration areas, eagle nests, and winter ranges for a variety of waterfowl species; the results are published in the *Penobscot Bay Conservation Plan.* By rating the numbers of different species that use given areas, they have defined significant wildlife habitat areas, which are plotted here on a satellite image of the region. *Credit: P. Conkling, T. Ongaro, Island Institute; data courtesy of Maine Department of Inland Fisheries and Wildlife*

locating potential population habitats during the winter months through a study of soil maps, topographic maps, and air photos. By focusing field efforts at these specific areas, cooperating ecologists and botanists tripled the number of known occurrences in New England in one field season.

This approach can work equally well for important and rare natural community types as well. Certain natural communities of plants and animals occupy certain types of landforms, or exhibit a visual signature on air photos and more sophisticated remote sensing imagery. The distinctive spacing of pitch pine in a pitch pine and scrub oak barren, and the concentric vegetation zones ringing outwash plain pondshore, are readily discernible from the air.

Our ability to observe the patterns of landscape diversity at all scales is magnified many times by the availability of satellite imagery. Now we can see entire landscapes at a glance and can see how the "vast unbroken expanse of forest" is actually full of structural diversity at many scales, offering a constantly shifting but steady availability of resources for plants and animals. Satellite images can thus help identify, locate, and quantify habitat diversity, especially for rare and endangered species that may be habitat specialists. These images are an invaluable tool in helping us catalog the distribution and abundance of important habitat types and enhance our ability to conserve the richness and diversity of the Gulf of Maine (fig. 10.11).

Forest Birds and Landscape Texture

The forests in the Gulf of Maine watershed are host to two categories of avian occupants, migrants and residents. Beginning in late March and intensifying through the spring, the forests can be inundated with songbirds passing through on their way from the neotropics to the boreal forests of Canada. By June, year-round resident birds such as chickadees, nuthatches, and siskins are forced to share the forests with a rich community of neotropical migrants, including redstarts, tanagers, over a dozen wood warblers, and a host of other species whose total numbers can exceed 100 species.

Insects are what lure these tropical birds away from the rich forests of Central America and the tranquillity of the Caribbean islands. With the increasing day length, out from the bark of trees and up from the ground and from fresh meltwaters emerges a cornucopia of grubs, beetles, mayflies, moths, butterflies, blackflies, and mosquitoes. If you enjoy seeing and hearing birds, you are indebted to the insects.

Not surprisingly, most of the forest birds are insectivores and are members of such guilds as the flycatchers, foliage-gleaners, gnatcatchers, and woodpeckers. Their days are spent in nearly constant search for insects to feed themselves and their young. The relative abundance of insects in a forest is probably directly related to the diversity and availability of habitat niches for insects. Predicting which forests will support the most species of birds is in many ways the same problem as predicting (or measuring) which forests provide insects with standing and moving water for reproduction, a mixture of standing and fallen dead vegetation for overwintering and feeding, and availability of a diversity of flowering plants.

Structural diversity, the size, shape, and number of layers (or physiognomy) of the forest, is equally important, because it provides more microniches and offers the opportunity for many more insects to distribute themselves spatially throughout the forest. All else being equal, a large, multilayered forest with a well-defined groundcover layer, a subcanopy, and a large, expansive canopy will support far more species of birds than a small forest with only a single layer. Physical diversity within a forest begets species diversity among the birds, small mammals, insects, and amphibians, because the physical spaces in a forest translate into greater ecological opportunities (or niches) for breeding and feeding. For example, an ovenbird builds a ground nest (in the shape of a miniature oven) only in forests with a thick groundcover layer. Whip-poor-wills, on the other hand, like an open understory with lots of leaves to camouflage them and their open ground nests. Winter wrens and hermit thrushes will nest only in the tangle formed by fallen trees. There is a greater correlation between bird diversity and physiographic diversity than between bird richness and plant species richness.

Thus, maximizing landscapes for forests birds translates into optimizing insect abundance and physical structure in forest lands.

Conserving Biodiversity

Especially at the natural community level, satellite imagery offers untapped potential for detecting the best representative ecosystems. This new technology offers even greater promise for assessing, planning, and implementing the landscape-scale conservation effort that will be necessary to conserve the biodiversity of the Gulf of Maine. Assuming that the "coarse filter" theory is correct, the questions are: how big should the conservation areas be? How close together should they be? And how many do we need?

Ecologists and conservation scientists generally agree that large size is important. The minimum threshold is usually considered to be "large enough to withstand or absorb the constant changes wrought by nature at all scales"—from tiny patches to catastrophic disturbance.

Equally important is how reserves fit together across the landscape: how do we allow for the anticipated need for species to shift with changing climate and find suitable places to live as they do? Conservation scientists therefore debate the concepts of "corridors" and "connectivity": not so much their validity, but rather how to express them within a system of reserves. For example, if such a system of reserves is established, how "connected" might two reserves be that would be 20 miles (30 kilometers) distant? For birds, these areas might be close enough. Even a small spring azure butterfly may, in time and by chance, find itself blown from one to the other, and thereby bring a new set of genes to the new area. For an individual salamander, these two areas might just as well be on opposite sides of the globe. But for salamander populations (and those of other sedentary animals and plants), which continually disperse and move over the landscape over time, the areas might easily see genetic interchange—an important measure of connectedness. For wide-ranging animals, and the many animals and plants that yearly disperse widely on foot, on the wing, and by wind, water, or "hitchhiking," the areas might see regular exchange of individuals or genes.

But what if the two areas are separated by a large river, a city, or an interstate highway? Satellite imagery, and the technologies of computer-assisted spatial information management, give us an unprecedented perspective of landscape diversity (fig. 10.12). We can now actually see from Boston to Saint John and see the physical and geographical relationships between ecosystems, waterways, mountains, cities, towns, farms, timberlands, and islands. This perspective is invaluable in determining where to place conservation areas, and how to mesh our critical need for conserving biological diversity with the important goals of living space and healthy local economies for us.

Conserving terrestrial diversity will require continued study of the relationships between genes, species, and natural communities. Fortunately, each year brings new insight into the requirements and behaviors of natural systems, and technological advances that enable us to assimilate and analyze the growing body of information and make informed conservation decisions. Our goal should be a well-placed system of conservation areas, each of which will be sufficiently large to accommodate all the expected and unexpected natural changes wrought by nature, and which together will harbor the best representation of all the plants, animals, and natural landscapes of the Gulf of Maine.

These are the challenges for the conservation movement as we enter the twenty-first century. How well we do in comprehending the nature of landscapes and conserving biological diversity will depend in large part on our aggressive and perceptive use of satellite imagery.

Vinalhaven

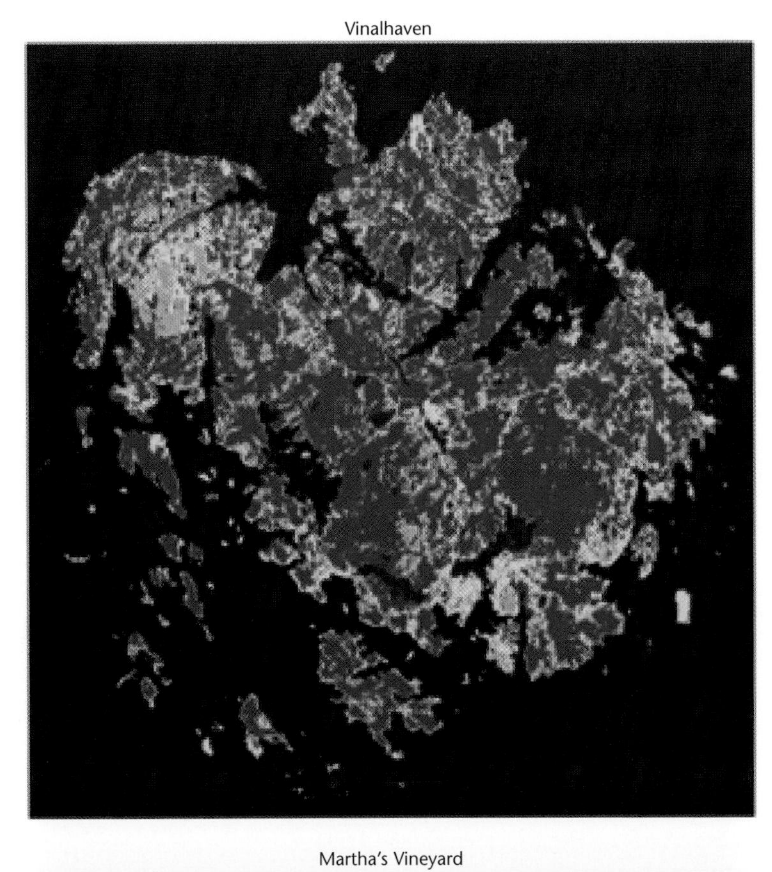

Grand Manan

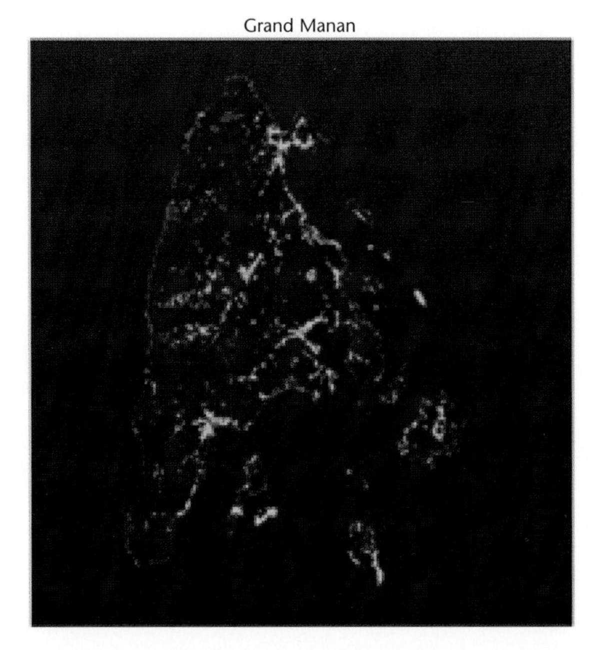

Nantucket

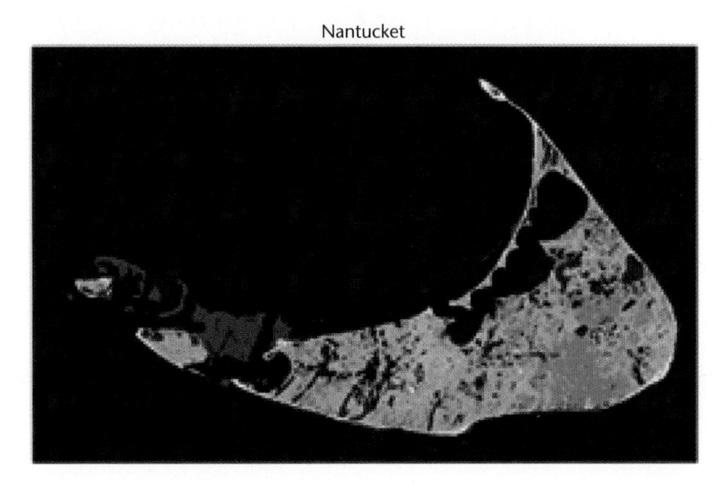

Martha's Vineyard

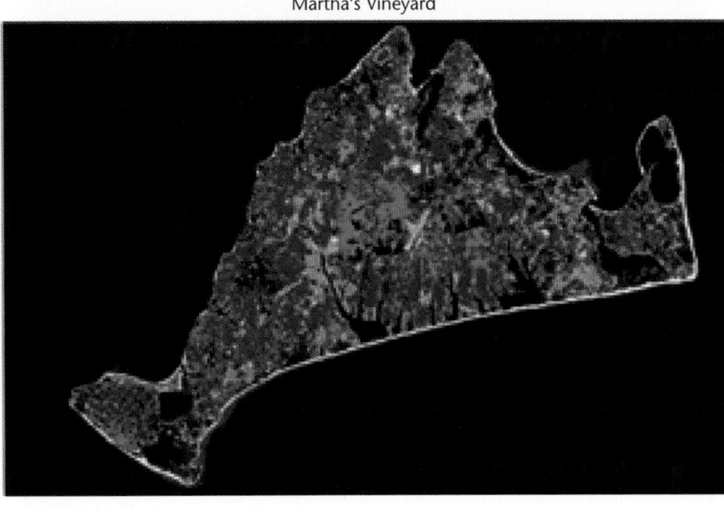

10.12 Habitat Diversity of Four Islands in the Gulf of Maine

Satellite image theme maps of the four large islands displayed here represent a transect across the region from southwest to northeast and inshore to offshore. At the southern end of the region, Martha's Vineyard and Nantucket enjoy a warmer climate, and therefore agricultural lands, grasslands, and oak forests are the prevalent habitats. But because Nantucket is farther out to sea (22 miles) than its neighbor, the percentage of heath/meadow land, which is adapted to more extreme wind and storms, is much higher than on Martha's Vineyard which is only 3 miles off the mainland. Midway around the rim of the Gulf, Vinalhaven, the largest island in Penobscot Bay, is densely forested with a mixture of softwood and hardwood forests and remains largely undeveloped. At the northeastern end of the Gulf, Grand Manan Island, which is surrounded by the cold Scotian Shelf waters, is covered almost completely by dense softwood forests, which are more frost-hardy than hardwoods. *Credit: P. Conkling, S. Meyer, T. Ongaro, Island Institute; GAIA image*

11.1 Matinicus Island Harbor

This nineteenth-century photograph shows the heart of the Matinicus Island fishing village, located 25 miles offshore where outer Penobscot Bay meets the Gulf of Maine. A flake yard for drying salt cod is a significant feature of the shoreland area used to cure fish from the prolific inshore groundfishery caught on handlines or trawls set all around the island. Men fished from small boats such as the pair of "Matinicus peapods" in the foreground, and a variety of other large and small craft moored in the harbor. Spruce cordwood stacked in the nearground was used for cooking and heat. The ledgy soil of the island, often grazed by sheep, suggests how the important resources for sustaining human life came from the sea, not from the land. *Credit: courtesy of P. Ralston*

11

HUMAN IMPACT

FROM SWORDFISH BONES TO SUSTAINABLE ECONOMIES

David D. Platt, Lloyd C. Irland, and Philip W. Conkling

The Gulf of Maine and its watershed have emerged into their present state from a dynamic relationship between people and land, beginning nearly 10,000 years ago when Native Americans first occupied the region. Sea level rise has obliterated many of the earliest coastal and island settlement sites, leaving their precise locations and character open to conjecture. Still, from the surviving evidence we know that the natural resources of the Gulf have profoundly shaped the way people lived, and that as the Gulf's waters have risen, cooled, or otherwise changed, species living in or around them have been forced to adapt. And with a changing climate, people have also been forced again and again to change their habits.

Ever since the coastline of the Gulf of Maine emerged from its covering of glacial ice, whales, fish, timber, granite, soils, and clean water have attracted people who wanted to use these resources. In all likelihood, natural resources will continue to be central features of economic life here, whether for present-day Native Americans or for other fishermen, foresters, and fish-farmers.

The Land Shapes Man: Shifting Diets

Swordfish backbones and ribs found at many archaeological sites around the Gulf of Maine indicate that this relatively warm-water fish once swam in inshore waters and that native people caught and ate them. About 3,800 years ago, however, rising sea level and the resulting increase in tidal range cooled the inshore waters, making them unattractive to swordfish. The bones stopped accumulating because native people were no longer catching the fish. They adapted: a changing environment had affected their diet.

Until the rising sea flooded the lower parts of the Damariscotta and New Meadows rivers in Maine (fig. 11.2), native people feasted regularly on freshwater oysters, leaving shell heaps some of which are visible today. Radiocarbon dating indicates that the most recent additions to these shell heaps occurred about 400 years ago, suggesting that oysters ceased to be part of the local diet at that time. Again, as the shape and

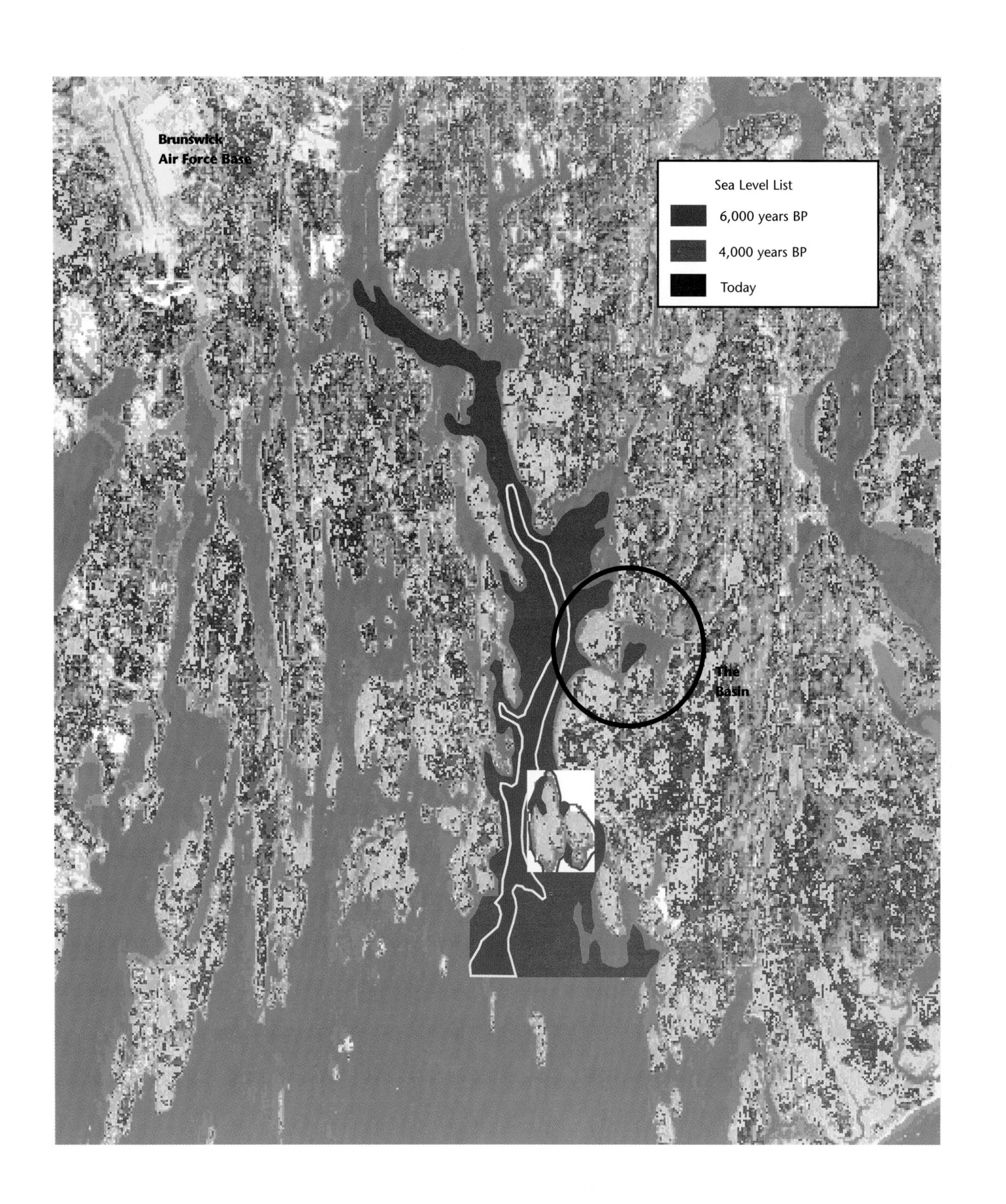
Brunswick
Air Force Base
Sea Level List
6,000 years BP
4,000 years BP
Today
The
Basin

11.2 Changing Shorelines of the New Meadows River

Casco Bay, in the Gulf of Maine, has been continuously occupied by humans for the last 11,000 years. During the past 6,000 years, rising sea level has affected the shoreline environments of the Basin, a fertile and productive hunting, fishing, and gathering ground for indigenous peoples. The excavation of Native American settlement sites provides a wealth of information that pertains both to the culture of the original inhabitants of the bay and to the environment in which they lived. Here sea level rise in the New Meadows River area has been plotted on a SPOT satellite image displaying near infrared spectral bands that clearly distinguish water bodies and wetlands from land mass.

In the summer of 1993, archaeological students from the University of Maine uncovered more than 10,000 artifacts at the Basin site, some dating back 8,000 years. They found fireplaces, shells, and bones piled in the water that once served as a breakwater and prevented beach erosion. It is estimated that 20 to 30 people lived at this site year-round. Dozens of other settlements dotted the coast as well, perhaps every half-mile or so.

The earliest inhabitants of the Maine coast enjoyed a rich and varied diet by exploiting the resources from several geographically remote microenvironments. Within a single year people may have been collecting shellfish on the tidal flats, hunting deer in the forests, trapping beaver in the streams, spearing and netting anadromous fish in the estuaries, and catching deep-water fish from stations on the outer islands of the bay.

Information on the environments of the past can be carefully reconstructed by calculating the age of fish and animal remains. Most of the fish excavated from the oldest sites were between 9 and 12 years old; some were as old as 19. They would have weighed between 30 and 60 pounds (15 and 30 kilograms), larger fish weighing over 100 pounds (50 kilograms). It appears that slightly older and larger fish were captured from 3,000 years ago up to the time of European contact. *Credit: graphic: S. Meyer, Island Institute; text: M. Bampton and N. Hamilton, departments of Geography and Anthropology, University of Southern Maine, Gorham*

ecology of the shoreline changed and the resource disappeared, harvesters were left with little choice and did without.

These adaptations were passive in character, of course, and they say more about the changing natural environment than they do about man's ability to change the landscape. But the arrival of Europeans in the region shortly after 1600 brought a new way of looking at things. Europeans came to fish, to trade, to extract resources for commercial purposes, and finally to settle and farm. Except for fishing, which in 1600 required few land-based resources beyond curing salt, space for drying fish, and small amounts of timber for buildings and boat repairs, the Europeans' activities were of the sort that would change the land itself. Inevitably, the age-old pattern would be reversed, and eventually even fishing—relatively benign in the seventeenth and eighteenth centuries—would become capable of widespread ecological destruction.

Man Shapes the Land: Lumbering, Clearing, and the Mast Trade

It would take 100 years or more for broad-scale changes such as land clearing to make a visible difference in much of the Gulf region, but it is logical to date their beginning from the arrival of the first European explorers. James Rosier's account of his 1605 visit to the Maine coast with George Waymouth and others makes repeated references to timber suitable for masts and the ease with which forests might be cleared for pasture—harbingers of the lumbering and farming that would eventually transform the region's landscape.

What sort of timber stood in the presettlement forest? Survey records from 1793 to 1827 compiled by forest ecologist Craig Lorimer for eastern Maine suggest a forest richer in hardwoods than it is today, with considerably less spruce and fir. Lorimer figured out the percentage of the trees of each species in the presettlement forest—20 percent spruce, 14 percent balsam fir, 12 percent cedar, 4.2 percent hemlock, 17 percent birch (two species), 15 percent beech, 8 percent maple (three species), and so on. Softwoods totaled 53.2 percent, hardwoods 46.7 percent.

Masts, a commodity of great military importance in the seventeenth century, were among the region's first important exports. The first cargo of mast pine left the Piscataqua River in 1634, and the trade accelerated after Great Britain was denied access to Baltic sources of masts in the 1660s. The Broad Arrow Policy in effect from 1691 until the American Revolution reserved all white pines over 24 inches in diameter on ungranted lands for the King. Mast agents were empowered to mark, cut, and ship masts to England. While mast agents were interested in only the largest and straightest pines and were not the cause of widespread land clearing, theirs was the first systematic effort to "high-grade" the forest and haul away its best trees. The roads and landings they built to remove huge trees must have opened up opportunities for further exploitation. At the same time, the presence of mast trees in some areas may have prevented some land clearing, as the trees' protected status meant they couldn't be cut without permission—a state of affairs resented by landowners anxious to sell timber, clear their acreage, and turn it into productive farmland.

The great rivers that drain this watershed—the Androscoggin, the Kennebec, the Penobscot, and the Saint John—allowed the mast agents to penetrate far into the interior seeking the tallest pines (fig. 11.3). Even today, the legal tradition continues: a stream is considered navigable if it will float a log. It is interesting to note that Europeans were not the first to make use of the region's large pine trees. Native people in Nova Scotia cut and hollowed pines of mastlike proportions for use as dugout canoes. The canoes were sufficiently seaworthy, according to reports, to transport warriors across the Bay of Fundy from Nova Scotia for raids on the Maine coast's early European settlements.

Far more trees fell for fuel and lumber than for masts or dugout canoes. The development of water-powered sawmills in the seventeenth and eighteenth centuries played a role, as did

11.3 Old-Growth Pines

White pines are among the longest-lived trees in the northern forest, occasionally reaching 400 years in age. This grove of old-growth pines located in Norway, Maine, is a registered Critical Area. *Credit: P. Conkling*

demand for fuel in Boston and other cities. Machias, settled in 1763, sawed a million and a half feet of lumber during the first winter of its existence. In two sketches of Machias done in 1776, a mere 13 years after the town was settled, there is not a tree in sight. Further west along the Maine and Massachusetts coasts, much of the land was already cleared and supporting pasture and annual crops of grain by this time. From a wilderness the first European settlers found foreign and frightening in 1600, two generations of lumbermen and farmers had begun creating an agricultural landscape that would last for more than two centuries. At the same time, they had set in motion a boom-and-bust pattern in resource use that would persist to the present day.

Shipping: Where Tree Cutting Was a Way of Life

Cribs, bedlogs, and breastworks of forgotten shipyards line the region's rivers to this day, a reminder that from the beginning to the end of the nineteenth century New England and the Canadian provinces surrounding the Gulf of Maine were home to a busy shipbuilding industry. It was a business that depended on ample supplies of wood, and from the start it played a role in the forest clearing that transformed the landscape.

"Much of the timber used was bought from neighboring farmers," writes William Hutchinson Rowe in his *Maritime History of Maine.* "Until well into the 1840's native woods were used almost exclusively." White ash, buttonwood, elm, hornbeam, ironwood, hackmatack, locust, maple, white oak, spruce, white pine, and yellow pine all went into ships in great quantities. Of this list, the only tree not dragged from New England forests was the southern yellow pine.

Descriptions of shipbuilding yards themselves suggest how much they must have affected the look of things. Rowe describes activities on the Kennebunk River in York County, Maine: "Here in the nineteenth century there were eight decades of shipbuilding . . . from 1800 to 1880 from six shipyards on this little river there came into being a total of 638 craft." Lack of deep water and the presence of two waterfalls on the Kennebunk necessitated the construction of a lock in the 1840s, a project that included a channel, massive granite walls, piling, ballasting, and great gates of hewn pine. Similar works, along with sawmills and other projects associated with shipbuilding and lumbering, went on all along the coast and up virtually all its rivers. Large-scale wooden shipbuilding continued, through good times and bad, until the last full-rigged wooden ship built in Maine, the *Aryan*, slid down the ways at Phippsburg in 1893.

From settlement to the late nineteenth century, millions of acres of forest were cleared for farming and settlement. By the middle of the nineteenth century, about 30 percent of Maine had been cleared and upwards of two-thirds of eastern Massachusetts had been transformed into an agricultural landscape. Meanwhile in the Canadian Maritimes places like the Annapolis River valley of Nova Scotia lay in open fields and pastures, and the forests to the northeast and west were being actively logged. Engravings, paintings, photographs, and maps of the settled parts of New England in the middle and latter years of the nineteenth century show a landscape largely tamed by man, far less forested than it is today (fig. 11.4). Stone walls, lilac bushes, cellar holes, abandoned roads, moss-covered stumps, and stone-lined wells speak today of a once-settled landscape, or at least a region of woodlots where tree cutting was a way of life.

Statistics, written descriptions, and pictures of the period also suggest something else: a society in which the depletion of natural resources, whether trees, farmland, rivers, or clean air, was accepted in the name of progress. It was a time when rivers all over the region were dammed for energy or for driving logs or put to use as sewers, with little heed to the effects on migrating fish. The forested region that had fulfilled the Royal Navy's need for spars in the eighteenth century had become a source of wealth for industrial America during the nineteenth and early twentieth century.

11.4 Harvard Forest Diorama

Ecologists from the Harvard Forest in Petersham, Massachusetts, near the New Hampshire border, have painstakingly researched the land use history of this section of northern New England. As part of their research, local ecologists have reconstructed a series of scenes from a single vantage point to illustrate the changes in the land over time. Three scenes are reproduced here, showing the shifting mosaic of forest and field from 1740, 1830, and 1915.

Records from Petersham and surrounding towns show how the region was gradually settled south to north between 1700 and 1760 as Indians were pushed out or bought out of their original territories. A settler would arrive in the early spring from some other area and begin clearing as much land as possible before winter set in. Tall, straight white pines and spruce were cut and set aside for sawing or hewing into building materials. Other trees were used to build a rough log cabin for immediate shelter, usually close to a spring or other water source. The rest of the trees were cut and burned as the clearing progressed, providing ash, or poor man's fertilizer for the new fields. Stumps were mostly left in place to rot, and it would be decades, if not generations, before all the stones were hauled out of the fields to build stone walls around the edges. A few large shade trees might be left in a new field or between clearings. But early on in each new town, settlers would set up a sawmill at a stream rapid or a tidal "privilege," a critical feature for attracting new settlers. The first sawmill in the New World was constructed in what would become Berwick, Maine, in 1623.

Sometime between 1830 and 1880, depending on the town, northern agriculture reached its maximum. The volume of stones that were hauled out of fields and the miles upon miles of stone walls that were built are an enduring testament to the hard work and industry of the region's settlers. Between two-thirds and three-quarters of the land in eastern Massachusetts, in coastal and central Maine, and along the shores of Atlantic Canada was cleared for crops, orchards, and pasture. The remainder of the forested land was subject to repeated cuttings for fuel, farm implements, and house and boat construction. Demand for wood to satisfy these needs was intense, and trees were viewed as another crop produced by the farm. It was not unusual for large farmhouses to burn ten or twelve cords of firewood during a winter.

Following the Civil War, farmers began leaving New England in droves, and many of these farms were abandoned agriculturally. The cleared land was quickly reforested; a large percentage of it grew into pure stands of white pine. Unlike the original stands of old-growth white pine used for masts and clapboards, the second-growth pine stands were full of crooked, knotty trees of low value. But they began maturing as manufacturing in the region reached its peak. Everything from shoes to textiles to apples to canned lobster needed to be packed and shipped to market, and thrifty farmers realized that old field white pine timber could be converted into cheap boxes. Soon a new industry sprang up as portable sawmills were set up on the "back forty" of thousands of farms throughout the region. Although of low quality, these second-growth stands were densely stocked and yielded between 25 and 50 thousand board feet of boxboards per acre. At a value to the farmer of $10 per thousand, a farmer with 100 acres of pine might be paid $30,000 to have an unplanted, seemingly worthless crop clear-cut. Between 1895 and 1925, Harvard foresters estimate 15 billion board feet of second-growth pine were cut in the region, with a manufactured value of $400,000,000!

Although few farmers or loggers probably noticed at the time, these white pine stands supported another crop of shade-loving hardwoods underneath them. By clear-cutting these stands in strips with horse crews and by piling the slash in windrows, the loggers released a new stand of hardwoods. Hardwood forests still dominate a great deal of similar land throughout northern New England and New Brunswick. *Credit: photos and data courtesy of Harvard Forest, Petersham, Massachusetts; text: P. Conkling*

Reforestation and Suburban Sprawl

Land clearing seems to have reached a peak in the Gulf of Maine watershed between the 1850s and the 1880s, except in the potato country along the upper Saint John River, where the peak did not come until the late 1940s. By the late nineteenth century railroads had opened western lands for settlement, and the development of agricultural machinery made small-scale farming uncompetitive. In the early twentieth century, the development of gasoline power rendered millions more acres of eastern farmland unnecessary for producing fodder for horses. Suddenly the family farms that had been shaping the landscape since the seventeenth century were doomed.

Since the beginning of the twentieth century, the dominant trends affecting Gulf of Maine landscapes have been the seemingly countervalent ones of reforestation and the sprawl of cities and suburbs across second-growth woodlots and abandoned farmland. The spreading cities and suburbs make some of the most compelling images. A comparison of aerial photographs taken 22 years apart just north of Bangor, for example, shows what modern mall development does to a landscape, replacing fields and farms with parking lots, roads, and buildings. Driving this sort of development all over the United States was the construction of interstate highways, including I-95, which passes alongside the Bangor Mall and transformed the farmland in that area into potentially profitable commercial real estate (fig. 11.5).

Urban land uses in the Gulf of Maine region remain small in comparison with other places in the United States and Canada, accounting for only a few percent of the total land area. The long history of ownership means that rural land is often owned in small parcels, and efforts to control or regulate land uses must deal with the plans, financial interests, and sentiments of large numbers of people.

11.5 Bangor Mall Development

Aerial photos are useful in measuring land use changes over small geographic areas. The 1971 photograph, at left, shows the site and surrounding land patterns prior to the construction of the Bangor Mall. The aerial on the right shows the same area in mid-1993. Land planners trace areas of growth and, by digitizing them, measure land cover changes over time. An aerial from 1951 would show this area without an interstate highway and as primarily agricultural land. *Credit: James Sewall Co.*

Wetlands and Estuaries

For 300 years, wetlands, long considered useless for agriculture, forestry, housing, or other humans uses, were drained, filled, diked, and otherwise altered (fig. 11.6). The sheer physical change was enormous, particularly in settled areas such as Portland or Boston where entire waterfronts and neighborhoods—Commercial Street in Portland and Back Bay and Logan Airport in Boston are typical—were created atop fill dumped directly into swamps, marshes, or shallow water. In areas where rusticating and tourism were important, early mosquito control measures significantly altered the salt marshes: ditching to drain small pools rearranged the natural drainage, and later annual spraying of DDT left behind pesticide residues that are still there.

Occasionally, altered wetlands have proved scientifically useful. The hay dikes in Machias are an excellent gauge of changing sea levels in eastern Maine, where a rising ocean and a coastline warping downward combine to produce some of the fastest rates of sea level change in the region. The age and original height of the dikes is known; the fact that tides overtop them today, flooding the former hay marshes once again with salt water, makes it possible to calculate the rate of sea level rise over the past century and a half with great accuracy. Old cribwork in the area provides further documentation.

As discussed in chapter 8, estuaries, where fresh and salt water mix, are areas in the Gulf of Maine where human activity is having its greatest impact. Casco Bay estuary, for instance, is believed by the National Oceanic and Atmospheric Administration (NOAA) and the Environmental Protection Agency (EPA) to be concentrating a variety of toxic dissolved substances. Similarly, Narraguagus Bay, located at the mouth of the Narraguagus River at Milbridge, is believed to be concentrating a variety of dissolved substances from agricultural runoff. The reason for this lies in the land use practices in the estuarine drainage area. The drainage area for the Narraguagus estuary is 416 square miles, of which 83 square miles, or about 20 percent, are in agriculture and range land. One hundred and six tons of nitrogen and 12 tons of phosphorus generated by agricultural uses enter Narraguagus Bay annually.

Understanding the connections between activities on the land and the quality of the water in bays and estuaries has been slow and difficult. Chesapeake Bay watermen learned it the hard way, when massive industrial development upriver destroyed their fisheries. The Gulf of Maine is larger and less industrially besieged, but learning is still taking place. Excluding inner Casco Bay, the mouth of the Piscataqua River, and Boston and Saint John harbors, much of the Gulf of Maine has time to act on the lessons learned elsewhere.

A Tale of Two Harbors: Boston and Saint John

In 1995, the Massachusetts Water Resources Authority begins operating a new sewage discharge tunnel running nine miles offshore (fig. 11.7). The tunnel will carry treated wastewater away from Boston, making it possible to clean up Boston Harbor and the coastline to the city's north and south for the first time in decades. The positive effects are easily cataloged: moving the Deer Island treatment plant's outfall from a site where the ocean is 30 feet deep and half a mile from shore to a new site nine miles offshore where the water is 100 feet deep will greatly increase dilution while keeping effluent away from beaches and shellfish beds. The new outfall will be located "where it will provide sufficient dilution of the effluent to protect public health and aquatic life," maintains the Massachusetts Water Resources Authority, the tunnel's builder (fig. 11.8).

But like any large event, natural or man-made, the construction of Boston's sewage tunnel can be expected to change things. Moving the point of discharge offshore will put it where water quality is now less affected by human activity. By adding nitrogen to the seawater, the daily discharge of 450 million gallons (1.7 million cubic meters) of treated sewage could con-

11.6 Diking in Acadian Salt Marshes of New Brunswick

The many wetland environments of the Fundy region—mudflats, salt marsh, and freshwater marsh—are of national and international importance as waterfowl breeding and staging habitat, and they form a critical link in the international network of habitats for migrating shorebirds. But the salt marshes of the upper Bay of Fundy have shrunk dramatically since the arrival of Europeans in the seventeenth century. Over the intervening years, settlers diked and drained 90 percent of the original salt marsh for agriculture, primarily hay production.

In the last half of this century, however, there has been a decline in agricultural use of the dikeland. Some of the more fertile dikeland, closest to the dikes and sea, lies idle, while freshwater wetlands are being created on more poorly drained dikeland soils farther from the coast in an attempt to recover habitat for wildlife. This region represents the most managed landscape along the coast of the Gulf of Maine. It took the tides of Fundy more than 3,000 years to create the coastal wetlands of Chignecto; it took barely 300 years to convert those wetlands into a largely engineered environment, and less than 30 to redo the engineering using modern machinery and methods. Natural cycles and processes continue to operate here, but they now do so in ways that are limited by, and subordinate to, human intervention. Without maintenance of the dikes, the area would revert to salt marsh. *Credit: text adapted from The State of Canada's Environment, 1991, copyright Minister of Supply and Services Canada, 1991*

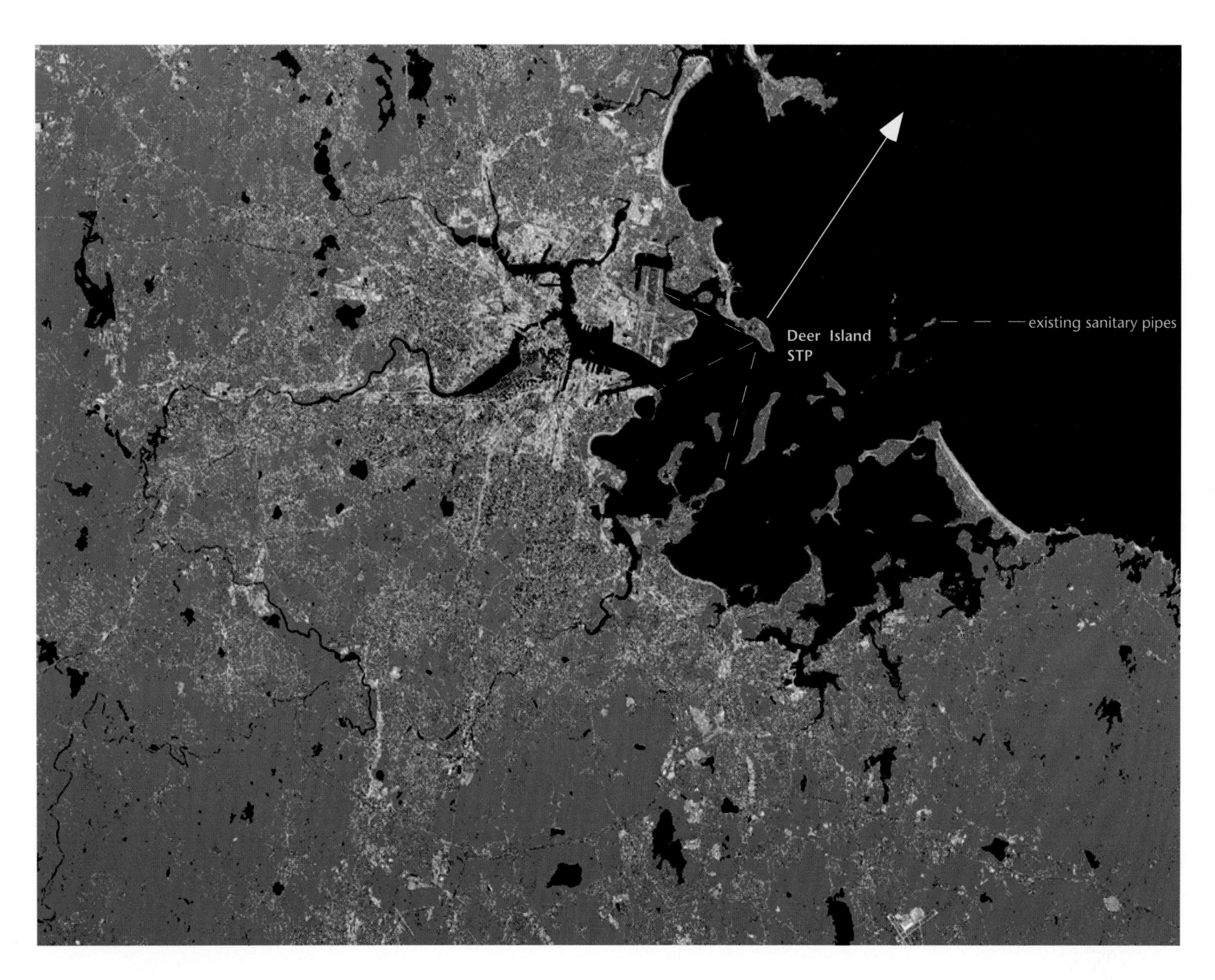

11.7 Boston Harbor Outfall

When construction is completed, the outfall pipe from Boston Harbor's new sewage treatment plant at Deer Island, which collects sewage from the entire metropolitan area, will run nine and a half miles (15 kilometers) offshore, terminating 400 feet (120 meters) below sea level. *Credit: Island Institute, S. Meyer*

11.8 Boston Harbor Tunnel

This cross section of the bottom of Massachusetts Bay with the sewer outfall pipe suggests the engineering complexity associated with this project. The final cost of construction of the Boston Harbor sewage treatment system is projected to be $4.5 billion when it is completed in 1999. Some environmental questions remain, such as the impact the millions of gallons per day of effluent at the end of the pipe will have on the phytoplankton community. These questions are important because two endangered species, the right whale and humpback whale, feed directly on phytoplankton. Recent whale mortalities in the Gulf of Maine have been linked to toxic algal bloom outbreaks. *Credit: T. Christensen, from Massachusetts Water Resources Authority data*

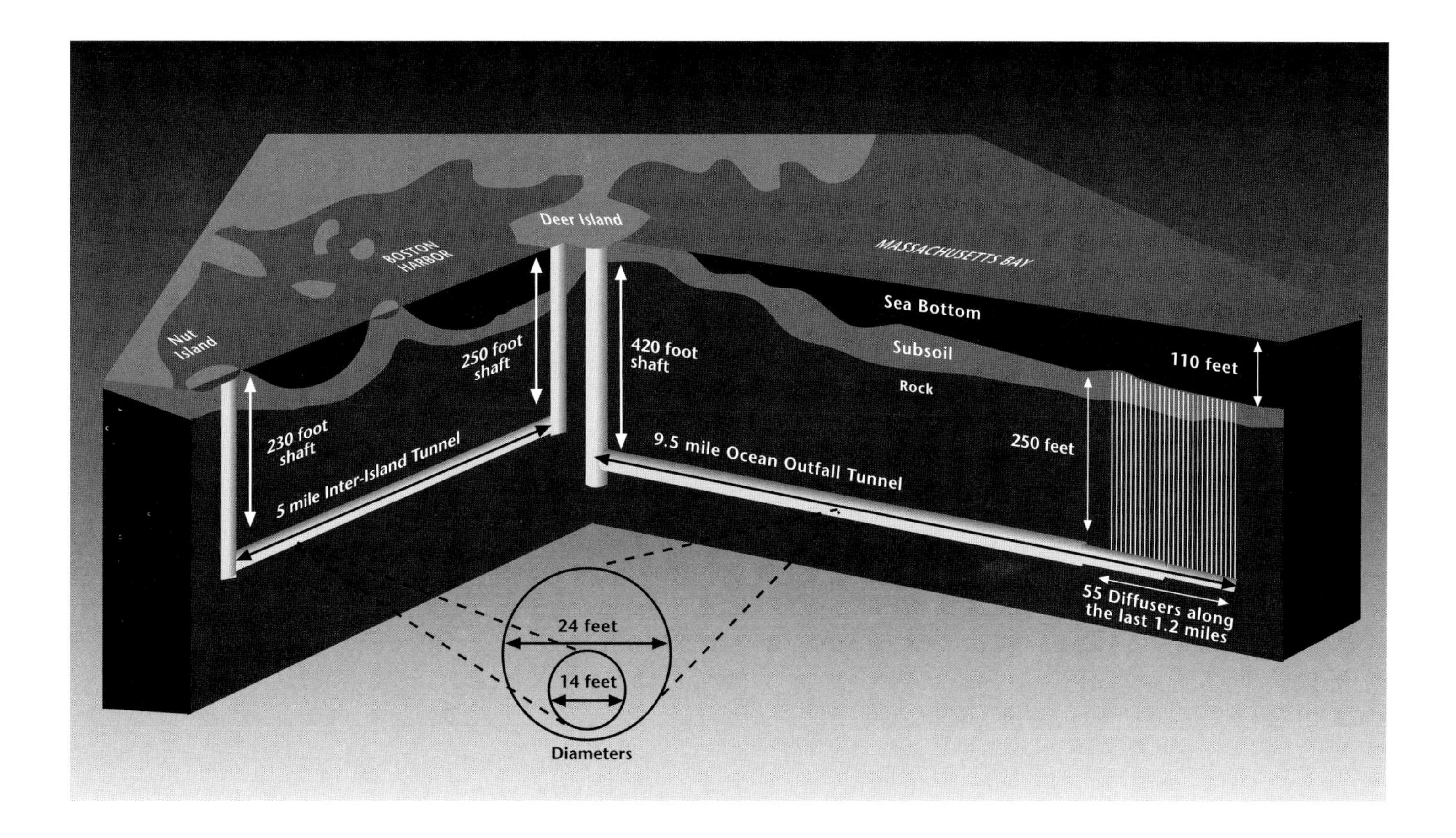

tribute to red tide blooms that in turn might affect the food supply of whales that feed on Stellwagen Bank and in sections of Cape Cod Bay. (In 1980, according to a group opposing the tunnel project, at least six Massachusetts Bay humpback whales were found dead with red tide–infested fish in their stomachs.) However, a U.S. Geological Survey report in 1992 concluded that the effects of the project wouldn't be serious enough to warrant stopping it, and the Water Resources Authority has agreed to modifications proposed by critics in hopes of keeping nutrients below a threshold that might trigger red tide blooms. A lawsuit filed in the fall of 1993 to stop the project suggests, however, that concern about its environmental impact remains.

At the other end of the Gulf of Maine, Saint John Harbour is on the receiving end of a cornucopia of pollutants making it the Bay of Fundy's most serious environmental hot spot. The Saint John River drains 55,143 square kilometers of land, flowing past three pulp and paper mills and various food processing plants and sewage treatment plants. All sewage from the central and south end of Saint John—55 percent of the total city sewage or 23,365 cubic meters per day—is dumped without any treatment at all. Another 25 percent is given only primary treatment to screen and settle out solids. Only 20 percent of the total receives secondary treatment where sewage is actually decomposed.

While the tremendous volume of water flowing into the harbor over the Reversing Falls makes it difficult to detect individual contaminants in the water, the tri-level governmental Saint John Harbour Study released in the summer of 1993 found that the water itself is toxic to the *Daphnia* water flea used in testing. Tests also found that water taken from both Marsh Creek and Little River, which also flow into the harbor, is toxic to *Daphnia* and to minnows. Levels of mercury contamination in striped bass in the Saint John River are almost five times higher than what is considered acceptable for human consumption.

The pollution flowing through Saint John Harbour has dramatically damaged the sea bottom. Murky water and sediment on the bottom have reduced the variety of marine life that formerly thrived there. Sediments contain toxic chemicals such as metals, PCBs, and PAHs, particularly in Courtenay Bay. These can enter the marine food chain and be carried to the rest of the Bay of Fundy in marine life. It should come as no surprise that some fish caught in the harbor area show fin rot, lesions, and other physical symptoms of stress from chemical exposure.

Other Invisible Changes

Heavy metals have found their way into the Gulf of Maine, largely via rivers in the region. Invisible to the naked eye, their presence can be detected by AVHRR satellites that sense surface water temperatures. The pulse of metals that flows periodically into the Gulf from the Kennebec River is one of the best examples of this phenomenon: images obtained and processed by scientists at the Bigelow Laboratory for Ocean Sciences in Boothbay Harbor, Maine, show a plume of contaminated sediment at the river's mouth. Snowmelt, heavy rains, or storm events upstream can cause the plume of sediments to expand as far south as Cape Elizabeth, Maine (fig. 11.9).

Non-point-source pollution shows up in satellite images of the Gulf of Maine (the same AVHRR images that reveal the presence of heavy metals) after heavy rainstorms or spring freshets, when it contributes to the plumes injected into the Gulf by swollen rivers. Not as visible but equally capable of causing harm is the pollution from non-point sources (particularly coliform bacteria from straight pipes and poor sanitary systems) that contaminates beaches and commercial shellfishing areas. The Gulf of Maine Council on the Marine Environment has established a series of monitoring points in sensitive estuarine and nearshore marine environments to detect the presence of a variety of heavy metal and other toxins. Blue mussel sites are selected for testing because they feed by filtering water through their gut and are therefore sensitive biological indicators for the whole marine environment (fig. 11.10).

11.9 Kennebec River Plume

Rivers of the Gulf of Maine contribute the greatest amount of fresh water during the spring. In the spring flood of April 1987, an unusually high volume of fresh water was released from the Kennebec (K) and Androscoggin (A) rivers. In highly turbid water, fine-grained sediments are suspended (a primary transport mechanism for toxic chemicals), changing the color of the water as far as it reaches. An AVHRR satellite recorded this turbid water and showed its areal extent. The sediment plume (in red and yellow) is shown in this series of satellite images. Note the back eddy of the plume into Casco Bay (Portland, Maine) in the April 2 image. *Credit: courtesy of U.S. Geological Survey*

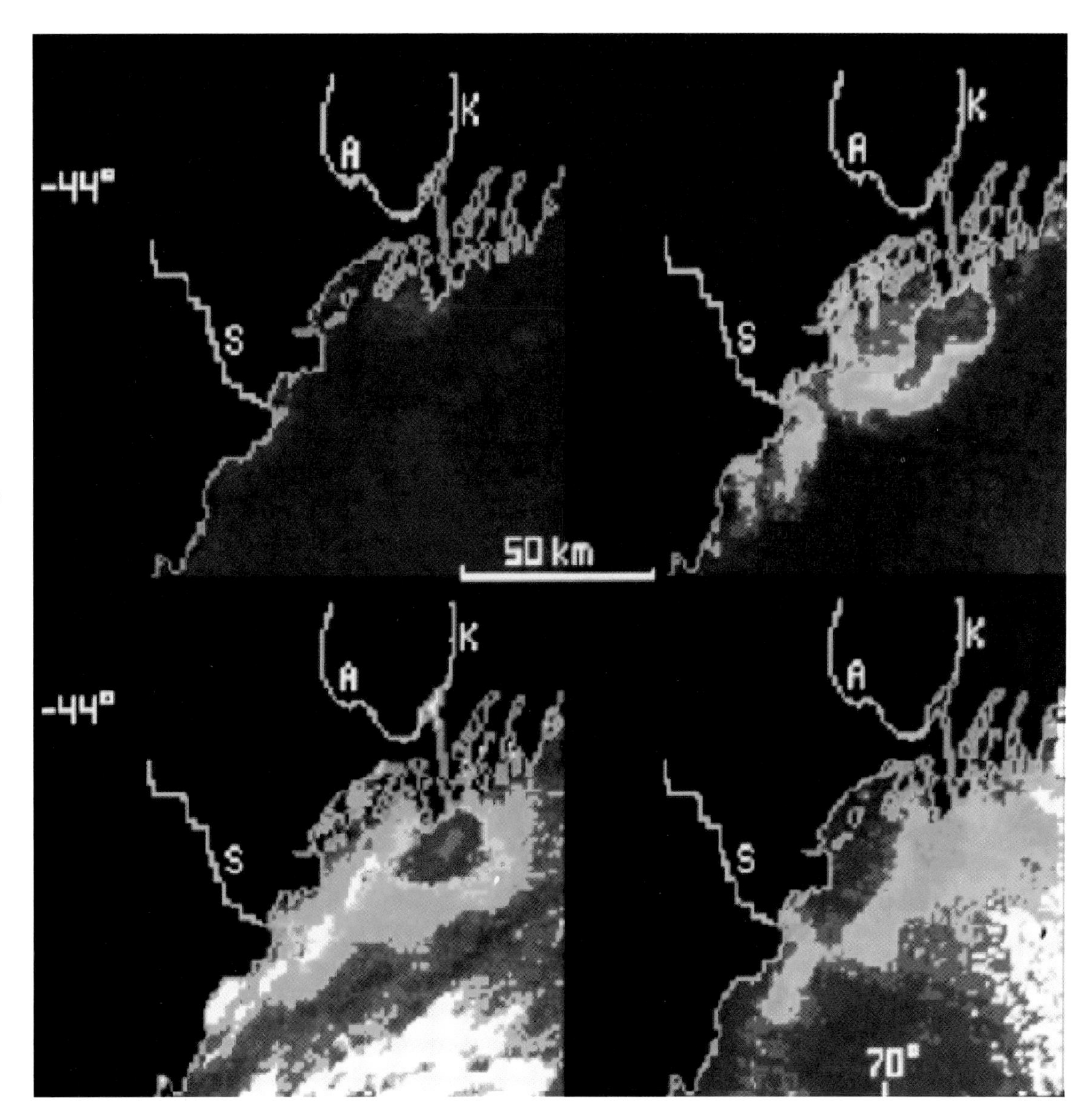

11.10 Blue Mussel Monitoring Sites

Blue mussels (*Mytilus edulis*), an indigenous species in the Gulf of Maine region, are being used as a bioindicator of contaminants at 33 sites throughout the region. Due to the nature of their water-filtering mechanisms, various mussel species are used internationally to measure heavy metals and chemicals in the nearshore environment in the estuarine discharge zones of different river systems. In the first year of the Gulfwatch Blue Mussel Monitoring Program, early indications found that field-collected mussels had been exposed to cadmium, copper, mercury, silver, chromium, lead, nickel, and zinc to different extents. Four metals in particular that are indicative of anthropogenic activities, silver, copper, chromium, and mercury, are found in higher concentrations in Boston and New Hampshire than further north.

The Gulfwatch program reports annually its analyses for toxic heavy metals and organic compounds such as pesticides and hydrocarbons. The report on toxic conditions is distributed to a wide array of public and private parties who may be able to incorporate the information into their decision making.

Pink dots on the map represent seeded mussel beds. Dark green is the international Gulf of Maine watershed. Bathymetry, in shades of gray, is shown in 100-meter intervals. *Credit: S. Meyer; data courtesy of John Sowles, Maine Department of Environmental Protection, from Gulfwatch Overview, Gulf of Maine Council on the Marine Environment*

In the United States, regulations have recently been enacted to protect water from non-point-source pollution such as runoff from farm fields and parking lots. Maine, New Hampshire, and Massachusetts have all adopted management plans as required by the U.S. Clean Water Act. The rules list priority waters to be protected, identify important categories of non-point-source pollution, and define ways to deal with each category. In Canada, where the law does not expressly address non-point-source pollution, the federal government limits phosphorus in laundry detergents and prohibits certain products in pesticides. Nova Scotia and New Brunswick also regulate the use of pesticides on their own.

Wetlands regulations differ on the two sides of the border. In the United States, modern-day regulations adopted under the Clean Water Act are supposed to control the use of coastal and freshwater wetlands and prohibit construction on sand dunes in Maine, New Hampshire, and Massachusetts. The U.S. Army Corps of Engineers has a "no net loss" goal for wetlands in the United States designed to stop the filling that has already obliterated so many of these ecosystems. Canada doesn't protect wetlands as such, but development that threatens them may be reviewed under provincial laws. New Brunswick and Nova Scotia, in fact, both have laws *encouraging* the reclamation of wetlands for agricultural purposes.

Population Trends and Outlook

The Gulf of Maine's coastal region was one of the first focal points for European settlement of the continent, as a host of plaques on town greens and in parks reminds us. Today, however, the counties of the Gulf's watershed are home to a small proportion of the nearly 300 million residents of North America.

Only 3.6 million people live in these counties, most of them in the U.S. portion of the region. A large number of people visit the shores, forests, and waters of the region every year as tourists, supporting many communities and jobs. The population of this region reflects changing industrial fortunes over the generations. In Maine, for example, many coastal towns and islands have lost population since 1900, including all of Washington County. Since 1950, towns like Fort Kent on the Saint John and Mount Desert and Rockland on the coast have lost population. Suburbanized areas like Cape Elizabeth adjacent to Portland, the New Hampshire coast, and towns surrounding Boston and on Cape Cod have all gained (fig. 11.11). The watershed in New Brunswick and Nova Scotia is sparsely settled. From 1980 to 1986, many towns lost population in New Brunswick but many gained in Nova Scotia.

The outlook is for slow population growth in this region. Some resource-based towns will lose population while suburban areas will gain. The driving forces behind resource use are not the size of the present population, but rather the land use patterns, which respond to regional trends, and the ways in which resources are used and by which wastes are managed.

But one trend is certain: more and more people will settle along the fragile margins between land and water, whether that water is fresh or salt, lake or bay. This is a worldwide trend: in ecological terms, the human is a littoral (or shore-dwelling) species. We inhabit no other ecosystems more tenaciously than the edges between land and sea (fig. 11.12).

Toward a Definition of Sustainability

Human attitudes in the Gulf of Maine have ranged, over the past 10,000 years, from a reverence for land and sea to a belief that exploiting resources to the point of destruction is the inevitable price of progress. Within the last decade, a new set of ideas has emerged that can be summarized by the term sustainability, providing a fresh vantage point to measure our utilization of the natural resources on which the future of this region is so closely tied.

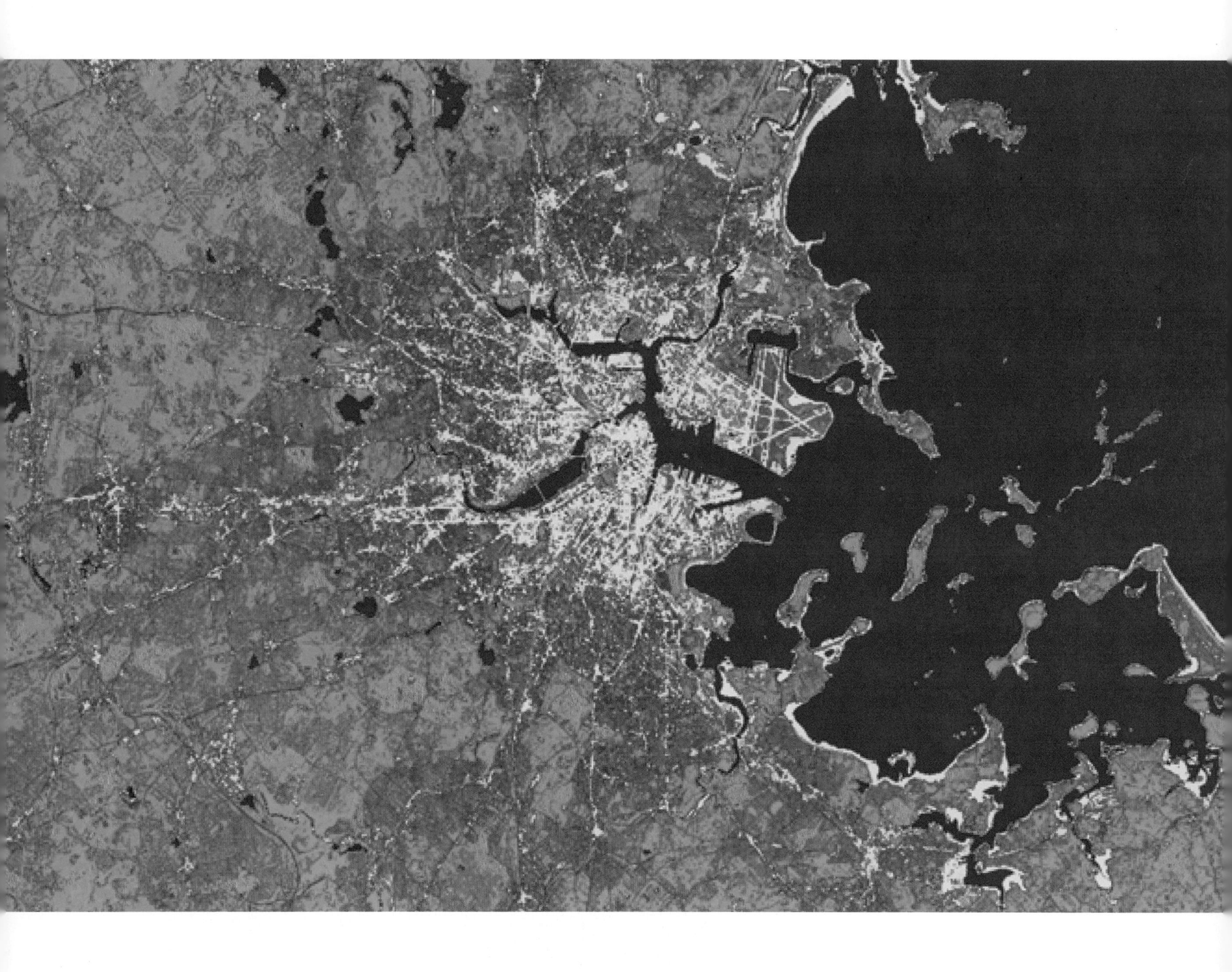

11.11 Human Patterns around the Boston Hub
(left)

The reason for Boston's nickname "the Hub" is suggested by this satellite image of the region. Many rivers and streams flow into the Boston Basin and ultimately into its harbor. The largest of these, the Charles, Mystic, and Neponset, converge on the harbor from the west, north, and southwest respectively. It is apparent that development (shown in yellow) has occurred most intensively along the shores of the area's watercourses, which were the earliest routes of transportation and commerce during colonial times. The patterns of human settlement that were established in an earlier era persist and are reinforced today by much larger transportation corridors, primarily highways, that are superimposed on older patterns. *Credit: P. Conkling, S. Meyer, Island Institute; GAIA image*

11.12 Human Patterns along the Shore
(right)

Ecologically, humans are a littoral species, occupying the shoreline environment more intensively and tenaciously than any other. *Credit: P. Ralston*

From the perspective of the past 400 years, it is evident that the particular resources that society is trying to sustain shift over time, often dramatically. In the eighteenth century, the King's ministers spent much time and treasure trying to sustain supplies of masts from the forests of New England and the Maritimes. Today, while large trees are valuable, there is no market for mast pines. As forest conditions and wood supply have changed, so also have demands.

There is a tendency to view sustainability as retaining the industry that is here now, as preserving "traditional ways of life." Yet if economic history tells us anything, it tells us that change is the way of life in the Gulf of Maine region. We no longer build wooden ships, load fuelwood and granite onto schooners for Boston, or pack lobsters in cans. Efforts to freeze the past are doomed to failure; what counts for local communities is the ability to adapt.

Sustainability is a complex concept. To apply it in practice, we have to consider how it relates to the economic base of local communities. We need to understand how our competitiveness relates to other resource-rich areas. But also we must begin to understand the ecosystem functions of our forests and waters. The addressing of these issues is at an early stage in this region (and worldwide). No doubt as the debate sharpens about what constitutes a "sustainable" practice, new issues and problems will arise. We will learn, in some cases, that what we are doing is ineffective or even counterproductive. We will also find that there are tradeoffs, and that there will always be a need to set priorities.

At the same time, we must be realistic. Citizens and governments cannot simply summon new twenty-first-century industries to appear from nowhere. Every other region in the world is trying to do that anyway. To sustain local economies will require a level of realism and detachment that is difficult to obtain. Today, the wood-based industries of Maine, for example, are more important to its manufacturing economy than they were in 1905. This must mean something. Economic activity based on the region's marine resources will continue to be important because large areas of the Gulf of Maine and Georges Bank boast a world-class marine productivity.

While a sound economic future for this region depends on sustaining the resource base of forests and waters, it depends on much more. Sustainability has particular meaning for regions such as the Gulf of Maine that are heavily dependent on fossil fuels, metals, and foodstuffs imported from elsewhere. There is often a tendency to take these supplies for granted, as if we could simply produce less of whatever seems inconvenient and buy it all elsewhere. But a region that takes sustainability seriously must make every effort to improve its self-sufficiency in many products, while at the same time it closes materials cycles and reduces its use of energy and materials. We must act as if materials are scarcer than they appear, given current markets. For example, renewable wood and wind energy appear expensive compared to surplus power we can buy from other places, but when we consider the environmental, foreign policy, and health costs of fossil fuels, it becomes clear that renewability has its merits.

The Gulf of Maine region has competitors in wood products, fisheries, tourism, high-tech manufacturing, and the service industries. As information becomes a global resource and the world economy more interlinked, the linkage between resource supplies and processing locations has loosened. There have never been so many choices available to those choosing locations for manufacturing plants, offices, or other job-creating activities. This fact threatens many traditional economic activities even as it beckons with new promises for the future. Noncompetitive regions will lose out.

11.13 Aroostook River Agricultural Lands

Aroostook County is famous for its potatoes from agricultural lands, shown here along the borders of the Aroostook River which flows north into the Saint John. The potato-producing towns of Masardis, Ashland, and Washburn are arranged along the river in this image from south to north. The topography of this region of Maine is of rolling hills cut by numerous streams and lakes. Squapan Lake is the large body of water at lower right. *Credit: P. Conkling, T. Ongaro, Island Institute; GAIA image*

A New and Finite World

A particular meaning of sustainability has come into focus recently: the idea of managing lands and fisheries to maintain and enhance ecosystem process and functions, instead of simply to maintain annual yields of products. Our present knowledge of ecosystems, deficient as it is, enables us to see ways we might improve things. Yet there is a danger in arguing that pristine, presettlement conditions should be the standard of comparison toward which all resource management should be directed. The past 400 years have seen the introduction of non-native diseases, the deletion of species, and significant changes in ecosystem structures. The climate itself has changed, perhaps in ways more significant than we realize, and the world's population is far larger than it was four centuries ago.

Perhaps the most important lesson for all citizens at the end of the twentieth century is that while hardly anyone noticed, all frontiers on the globe closed. There is simply no new place on earth to find unexploited forests or fisheries; no open space into which cities and suburbs can continue their historical expansion without affecting some aspect of local biodiversity; no refuge in the ocean where marine life is undisturbed; no stretch of shoreline or beach where human competition for natural resources is not already keen.

We are entering a time when everything we have come to accept as part of our civilization will be judged by new standards: we must make more out of less; we must accept the profoundly radical idea that all resources on which we depend are inherently limited; we must learn to think in cycles; we must appreciate and celebrate our part in an interconnected biological system; we must learn to accurately calculate what our products cost the environment. We face the awesome task of intervening in our communities' futures to insure that present trends of decline are arrested and reversed. But we do not face what so many other bioregions of the earth do: enormous restoration costs to repair badly damaged natural systems.

While it would be naive to think that such shifts in world view will be easy or painless, perhaps here in the Gulf of Maine we have a potential head start in coming to terms with a world of exquisitely finite resources. It is arguably easier here to see the deep ecological connections between forest uses, river quality, and marine life. The circular gyre of currents in the Gulf of Maine and the geomorphology of a sea within a sea are not just symbols of the region, but fundamental ecological conditions that serve to reinforce the critical idea of the structure of cycling and recycling mechanisms in nature. What goes around comes around; shifting environmental problems from one part of the biosphere to another—from land to water, from one side of an international commons to the other, from geologically based fossil fuels to the atmosphere—is not a sustainable practice.

Defining ecosystem health and function in this region's heavily used and modified forests and waters will be a major challenge, a learning process of historic proportions. A perspective on the region's resources such as we offer here can be helpful in leading governments, citizens, and members of the scientific community to a better appreciation of what must be done.

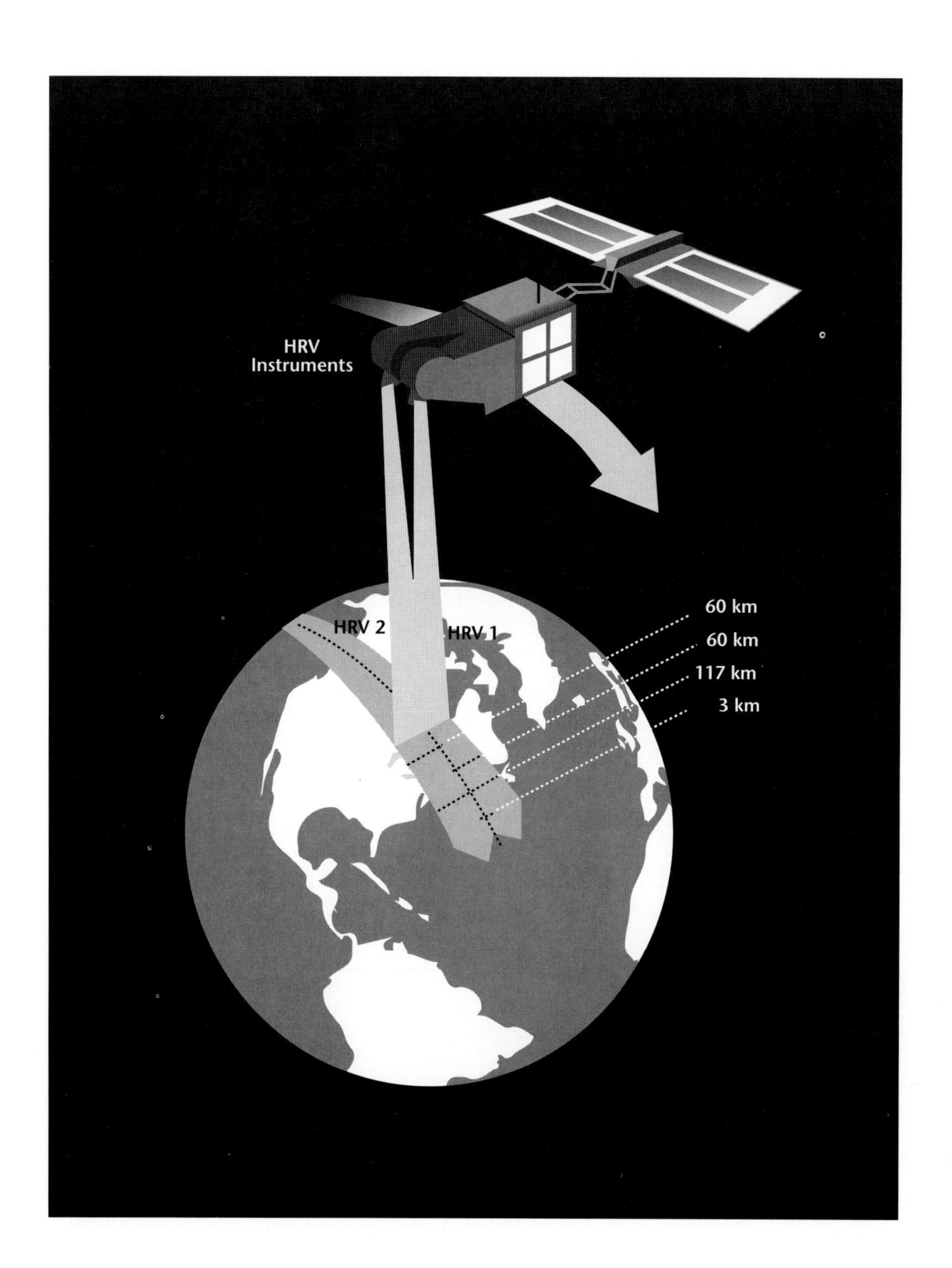

A.1 Satellite in Polar Orbit

A satellite in polar orbit (also referred to as a sun-synchronous orbit) circles the globe from pole to pole as the earth rotates underneath it; in this way coverage of every portion of the earth's surface is possible. Different sensors on board a given satellite are designed to capture various swath widths of differing spatial resolution as they orbit. *Credit: T. Christensen*

APPENDIX
UNDERSTANDING IMAGES OF THE EARTH

A PRIMER ON REMOTE SENSING

Janet W. Campbell and Cynthia B. Erickson

The technology of making and interpreting measurements of the earth from space or airborne sensors is known as remote sensing. Prior to 1960, the camera was the most widely used remote sensing tool, although infrared and radar systems had been developed and used during World War II. Space-based remote sensing began in 1960 with the launch of the first Television Infrared Observation Satellite (TIROS-1). The TIROS series of satellites were the predecessors of the NOAA polar-orbiting weather satellites currently operated by the National Oceanic and Atmospheric Administration (NOAA). The first land-oriented satellite was launched by NASA in July 1972. Named Landsat, it became the first of a series of six Landsat satellites designed to map features of the land surface. Today, there are dozens of environmental satellites launched and operated by various countries.

The scientific techniques and technological innovations that have emerged from the past 30 years of analyzing satellite images have fundamentally altered our view of the planet. For many years this technology was restricted to military applications or to research programs at big academic and commercial institutions. But recently the power of desktop computing and the proliferation of satellites from other countries have opened this frontier to people everywhere.

Images of the Gulf of Maine and its watershed derived from remote sensing techniques appear throughout this book. Some images look like photographs, and indeed some are; others look like abstract paintings. Most remote sensing images are not photographs but rather digital images produced by computers. Some of the important concepts involved in remote sensing are presented in this appendix.

Mapping the Earth's Surface

A wide range of remote sensing techniques exist for mapping

the earth's environment. Virtually all techniques employ sensors that record radiation traveling from the earth outward into space, but they differ in the type of radiation detected. Each type of radiation (reflected solar, thermal infrared, microwave, and so forth) is useful for asking different questions, as we discuss below.

Understanding spatial resolution is fundamental to all remote sensing techniques. A technique's resolution is defined by the smallest object that the sensor can detect. Land-oriented applications require much higher spatial resolution than other applications, because features of interest on land, such as cultivated fields, roads, and rivers, tend to be much smaller than those in the atmosphere and ocean. Sensors aboard the present Landsat satellites resolve features that are 30 meters in scale (approximately 100 feet), and sensors aboard the French Système Probatoire d'Observation de la Terre (SPOT) satellites resolve features that are 10–20 meters (30–60 feet) in scale. In contrast, the spatial resolution of oceanographic and meteorological satellite sensors is typically 1,000 meters or larger.

Another important aspect of satellite remote sensing is the frequency of coverage. This is largely determined by the orbit in which the satellite is placed. Weather satellites require the most frequent coverage. To observe the movement of weather systems, one needs to monitor changes occurring on times scales of minutes to hours. This is achieved by placing the satellite into a geostationary orbit. NOAA's Geostationary Operational Environmental Satellites (GOES) series and the European Space Agency's Meteosat satellites are an array of geostationary satellites positioned far above the equator, where they orbit at the same rate that the earth rotates. Even though they are actually moving much faster than other satellites, they appear to be stationary or "parked" over the same position on the earth. From a very high altitude (35,600 kilometers), their sensors are able to sweep the full earth disk—nearly an entire hemisphere—facing the satellite. Updated images of cloud patterns are provided every half-hour. These are the images used to produce film loops for television weather broadcasts.

Landsat and SPOT satellites, as well as the NOAA satellites, are in polar orbits at much lower altitudes (800–900 kilometers). Circling the earth from pole to pole, these are referred to as either polar or sun-synchronous orbits, the latter term because the satellite passes over each latitude at the same local time of day (fig. A.1). Sun angles, shadows, and other such effects remain similar or vary slowly in a predictable way. Most of the images seen in this book were obtained from polar-orbiting satellites.

As a satellite passes over an area, data are collected within a swath beneath it. Usually, the data are arranged along scan lines perpendicular to the satellite's ground track. Each scan line is divided into picture elements or pixels that have a finite area as determined by the sensor's spatial resolution (fig. A.2). For example, the Landsat satellite scans a swath 185 kilometers wide with a spatial resolution of 30 meters. Thus, there are over 6,000 pixels per line of Landsat data. When the lines of data are displayed side by side on a computer screen, the result is a satellite image. This digital image may resemble a photograph, but it is actually a display of pixels corresponding to the resolution of the sensor.

A.2 Pixels

Sensors aboard satellites detect reflectances on a grid of varying spatial resolution depending on the design of the individual instrument and its purpose. These data are then transmitted as a digital signal back to earth and the grid is displayed as picture elements, or pixels. In this SPOT image of an island and ledges (with 20-meter pixel resolution), you can clearly see individual pixels that represent spruce and field communities on land, as well as different depths and conditions of the surrounding water. Note how pixels can be counted to generate secondary measurements such as width, perimeter, area, etc. *Credit: S. Meyer, Island Institute*

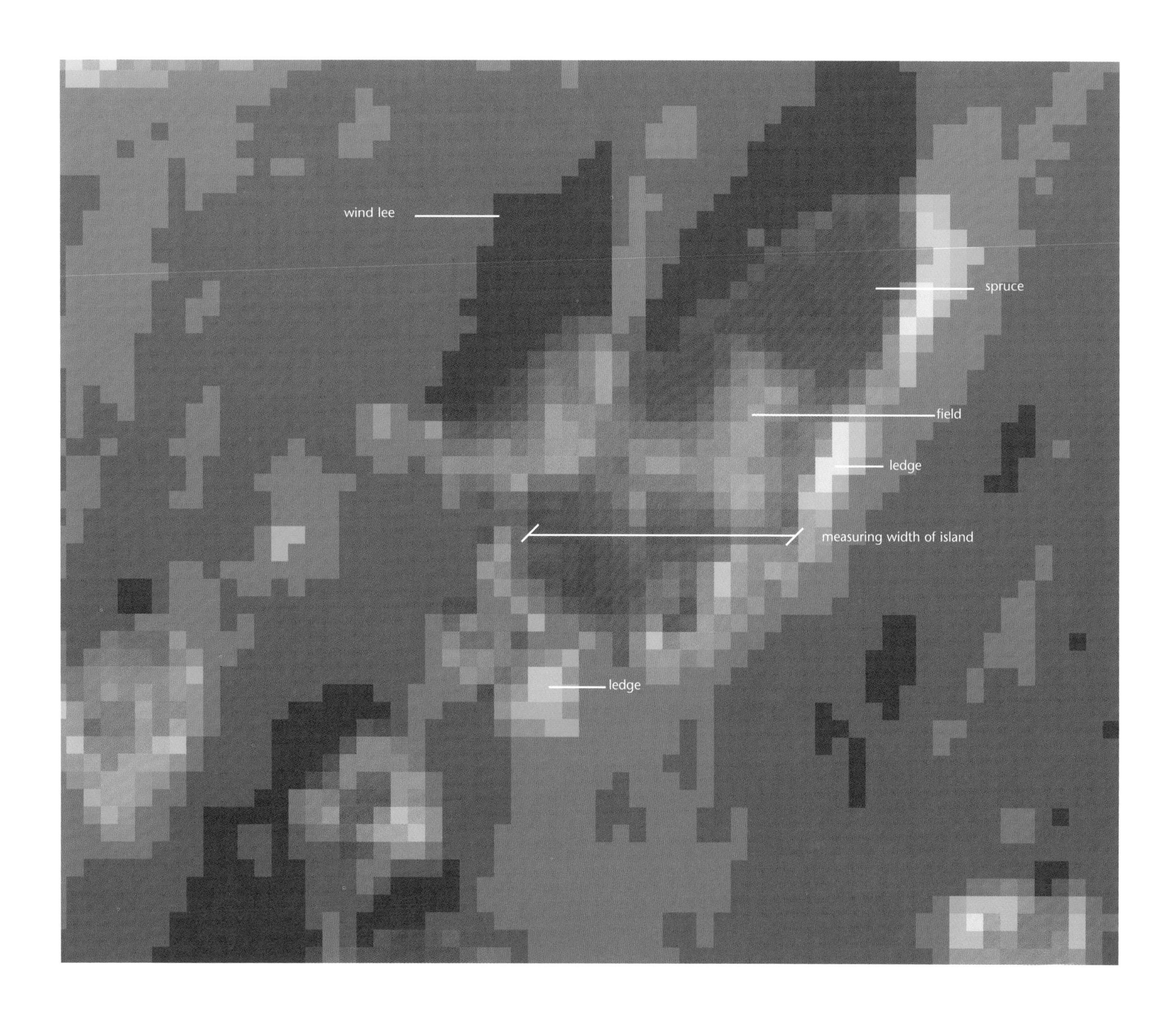
wind lee
spruce
field
ledge
measuring width of island
ledge

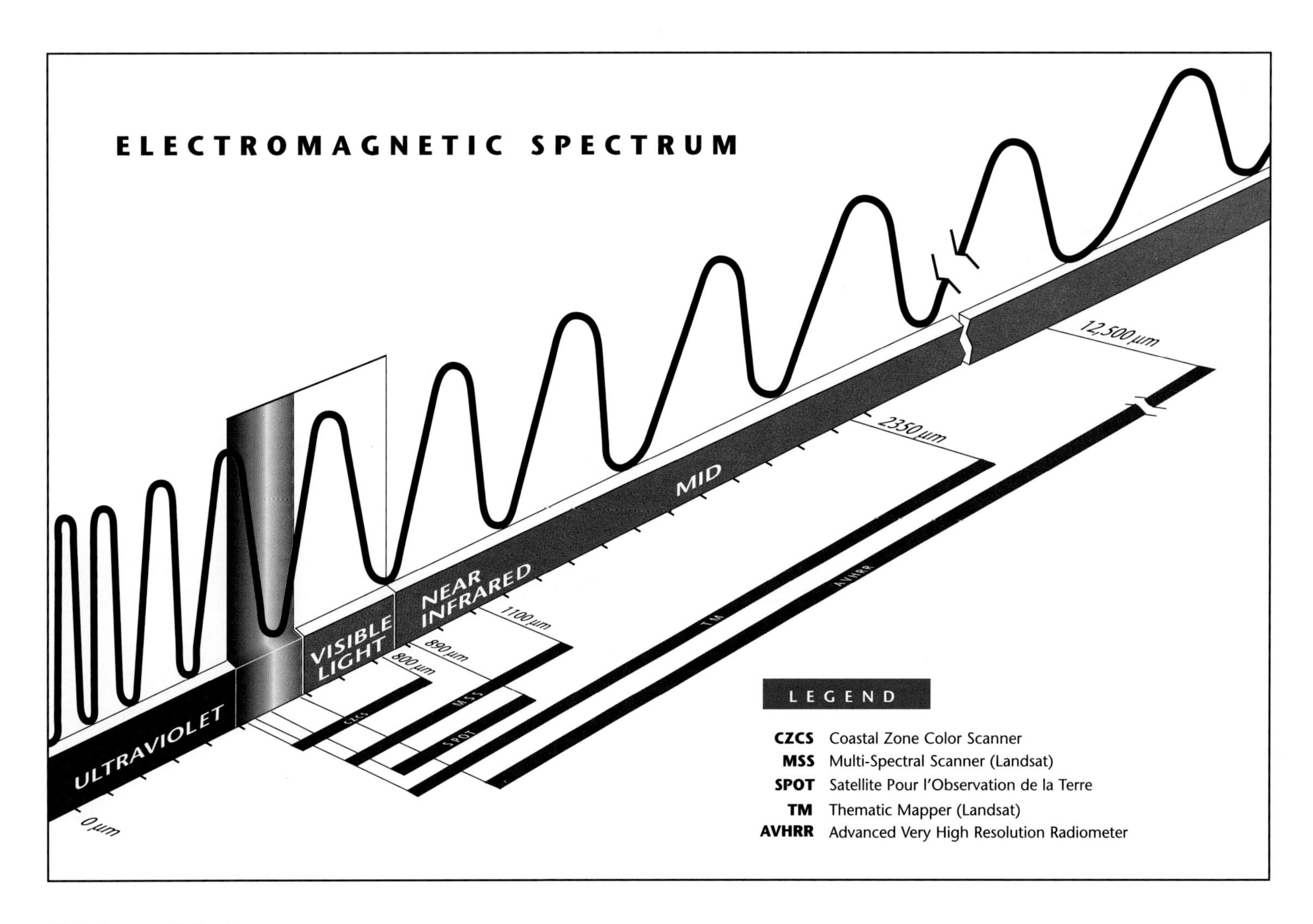

A.3 Electromagnetic Spectrum

The electromagnetic spectrum is divided into radiation of different wavelengths (the distance from crest to crest of a wave). Various sensors on board satellites are engineered to detect different wavelengths, not only in the visible part of the spectrum (e.g., Coastal Zone Color Scanner), but also in the near infrared and mid-infrared (e.g., SPOT and Landsat). *Credit: T. Christensen*

The Spectrum of Data

All satellite data originate as recordings of electromagnetic radiation detected by sensors aboard the satellite. The various remote sensing systems detect electromagnetic radiation from different spectral regions (fig. A.3). For example, Landsat and SPOT sensors detect solar radiation reflected back into space by clouds, land, and ocean surfaces; Landsat also has a sensor for detecting infrared radiation emitted by the earth and its atmosphere, and certain oceanographic satellites record "active" microwave (radar) signals sent by the sensor and reflected back by ocean waves.

Each spectral region conveys a unique set of information about the earth's environment. For example, since ozone absorbs ultraviolet radiation, measurements of ultraviolet radiation are used to map ozone in the atmosphere. Microwave techniques are used to monitor the polar ice sheets in an effort to detect melting that might accompany global warming. Visible and infrared data are used to assess the health of crops, forests, and other forms of vegetative cover. Thermal infrared radiation is used to determine the temperature of the land, clouds, and ocean surface.

Most of the satellite images in this book have been derived from reflected solar radiation. The sun's radiation includes visible light, which is composed of a mixture of colors ranging from violet to red. Each color is associated with a different wavelength, with violet having the shortest wavelength and red the longest. The sun's radiation also includes ultraviolet (UV) light at wavelengths shorter than visible light, and near infrared (NIR) and middle infrared (MIR) radiation at wavelengths longer than visible light.

Although there are panchromatic sensors that incorporate all visible light into one measurement (like black and white film), most visible-range sensors split the incoming radiation into different colors or spectral bands. They may also record various spectral bands of UV, NIR, or MIR radiation. Such data are called multispectral. For each pixel or resolution element in the satellite image, there may be as many as seven different spectral bands. There are hyperspectral remote sensors now being tested on aircraft for future satellite missions that have hundreds of spectral bands.

Spectral Signatures

To understand the nature of multispectral data, it is helpful to consider first how humans perceive color. Our eyes are highly sophisticated remote sensors capable of detecting numerous colors (spectral bands) of visible light. When sunlight falls on an object, some of the colors are absorbed and others are reflected. An object that appears red, for example, is reflecting red light while absorbing all other colors. The ocean appears blue because most of the light entering the ocean is absorbed, while only the blue light is reflected. If all incident light is reflected, the object appears white, whereas if all the light is absorbed the object appears black.

The key to interpreting multispectral data is understanding the reflectance properties of different surfaces or objects viewed by the sensor. The tendency of an object to reflect or absorb solar radiation at different wavelengths gives rise to its "spectral signature." We can predict the spectral signatures of objects within the range of visible light, since this is the spectral region that we see. For example, we would predict the ocean to have a high reflectance in blue spectral bands, vegetation to have high reflectance in green bands, and so forth. But remote sensors are not limited to detecting only the visible range.

The Landsat and SPOT satellite sensors record multispectral data in visible and infrared spectral bands. The SPOT multispectral sensor has three spectral bands: two in the visible range (green and red) and one in the near infrared region. Landsat's Thematic Mapper records reflected solar radiation in six bands, and thermal infrared radiation emitted by the earth in a seventh band. Of the six bands of reflected radiation, three are in the visible (blue, green, and red) and the remainder are in the near and middle infrared regions.

Scientists have learned to interpret reflectance patterns outside the visible spectral region, and in many instances it is this invisible information that accounts for the power of multispectral images. Near infrared radiation is almost completely absorbed by water, whereas land and particularly vegetation have high reflectances in the NIR region. Thus, the NIR bands are useful in differentiating land from water and wetlands. In addition, the NIR bands are useful in locating and identifying different species of vegetation, and in determining whether or not a particular plant species is healthy or diseased. Middle infrared bands are sensitive to moisture content and therefore are useful in vegetation studies and in locating clouds, snow, and ice formations. Thermal infrared bands measure the amount of heat emitted from an object. These can sometimes be calibrated to give the temperature of an object to within half a degree.

Alone or in conjunction with the IR bands, visible bands also tell a story about the land and water features of the earth. The blue and red bands are absorbed by the chlorophyll pigment present in plants, whereas the green band is reflected. These bands are therefore useful in determining plant species and the health of vegetation. Certain soil types and rock formations have enhanced red reflectances. Water has enhanced blue reflectance compared with other bands, but its major distinguishing characteristic is that is has a significantly lower reflectance in all visible and NIR bands compared with land and clouds. The visible bands are commonly used to locate clouds, which can be invisible in infrared bands if their temperature is the same as that of the land or ocean surface.

Interpreting Multispectral Data: Thematic Maps

Since features observed in satellite images generally have distinctive spectral signatures, multispectral data can be classified (clustered) to identify various spectral classes in the image. By this computer technique, all pixels with a similar spectral signature are assigned to the same class. This process reduces the multispectral data (multiple measurements for the same pixel) to a single value at each pixel. The value assigned or class number is usually arbitrary for a given image, but there is a need to group classes according to meaningful cover types, producing what is known as a thematic map or land use/land cover map.

For example, all pixels that represent forests may be grouped into a single "forest" category; divided into softwood (conifer) and hardwood (deciduous) forests; or further identified as white pine, white birch, and so forth. There is a limit to the specificity that can be achieved, because remote sensing information can be ambiguous. An asphalt parking lot can resemble the asphalt roof of a building; white sandy beaches are often indistinguishable from gravel pits. This calls for the need to ground-truth an image—the practice of verifying assignments made in a thematic map by visiting sites seen in the image (fig. A.4).

The process of classifying multispectral data and producing a thematic map is only one technique for interpreting multispectral satellite data. Other techniques involve mathematical formulas whereby data from two or more bands are used to derive a new "band." In the study of air pollution damage to white pines in New England, the ratio of Landsat bands 4 and 5 has been found to be an effective indicator of ozone damage. In oceanographic applications, the ratio of blue and green bands indicates the relative abundance of marine phytoplankton, and two thermal infrared bands are used to derive the sea surface temperature (see fig. 4.6).

Displaying the Data

Most computer screens are too small to display a full Landsat scene with over 6,000 pixels per line, or even a SPOT scene with 3,000 pixels per line. To display the full scene requires using subsamples of the data with only every *n*th pixel displayed,

A.4 Ground Truthing

Here an ecologist from the Island Institute (and editor of this book) is field-checking classifications of a satellite image in October 1989, before color laptop computers were available. Here, a desktop computer in the back of the pickup is wired into an inverter and then into the vehicle's electrical system to enable us to travel through the landscape, test our desk-truthing, and tease out difficult classifications. *Credit: R. Podolsky, Island Institute*

where *n* may range from 2 to 12 or more. Once the full scene has been displayed in this manner, one can zoom in to view a particular area in greater detail. Zooming in allows one to display a smaller portion of the image at higher resolution. Eventually, every screen pixel corresponds to one satellite image pixel. At this point, one can appreciate the spatial resolution of the data. Zooming in further enlarges the pixel so that each satellite pixel is displayed as a square area of uniform color. One cannot zoom in to see finer details beyond the spatial resolution of the sensor.

The colors or shades of gray in a remote sensing image depict the data collected by the satellite sensor. One technique, called a color composite, allows three spectral bands to be displayed simultaneously. The intensity of each band is represented by one of the computer display colors (red, green, or blue), and the combination of the three bands produces all possible colors. When the reds, greens, and blues of the computer display correspond to the same visible bands (red, green, and blue) measured by the satellite sensor, the result is called a true-color composite. A true-color image is intended to replicate what the human eye will see, and consequently the resulting image might resemble a real photograph. When the red-green-blue composite uses bands that differ from the red, green, and blue visible bands or includes additional spectral bands, the resulting image is called a false-color composite (fig. A.5). The image will often look very different from what the human eye would see. Finally, a palette of colors can also be arbitrarily assigned to data values, and when this is done the result may look highly abstract.

The spatial information extracted from a satellite image may become part of a geographic information system. As such, it becomes a layer of information that can be accessed with other layers such as topographic contours, geopolitical boundaries, property ownership maps, and so forth. The layers are related by the fact that they pertain to the same geographic base map (see fig. 1.7).

The concept of analyzing landscapes by means of overlays of different landscape attributes was first articulated by

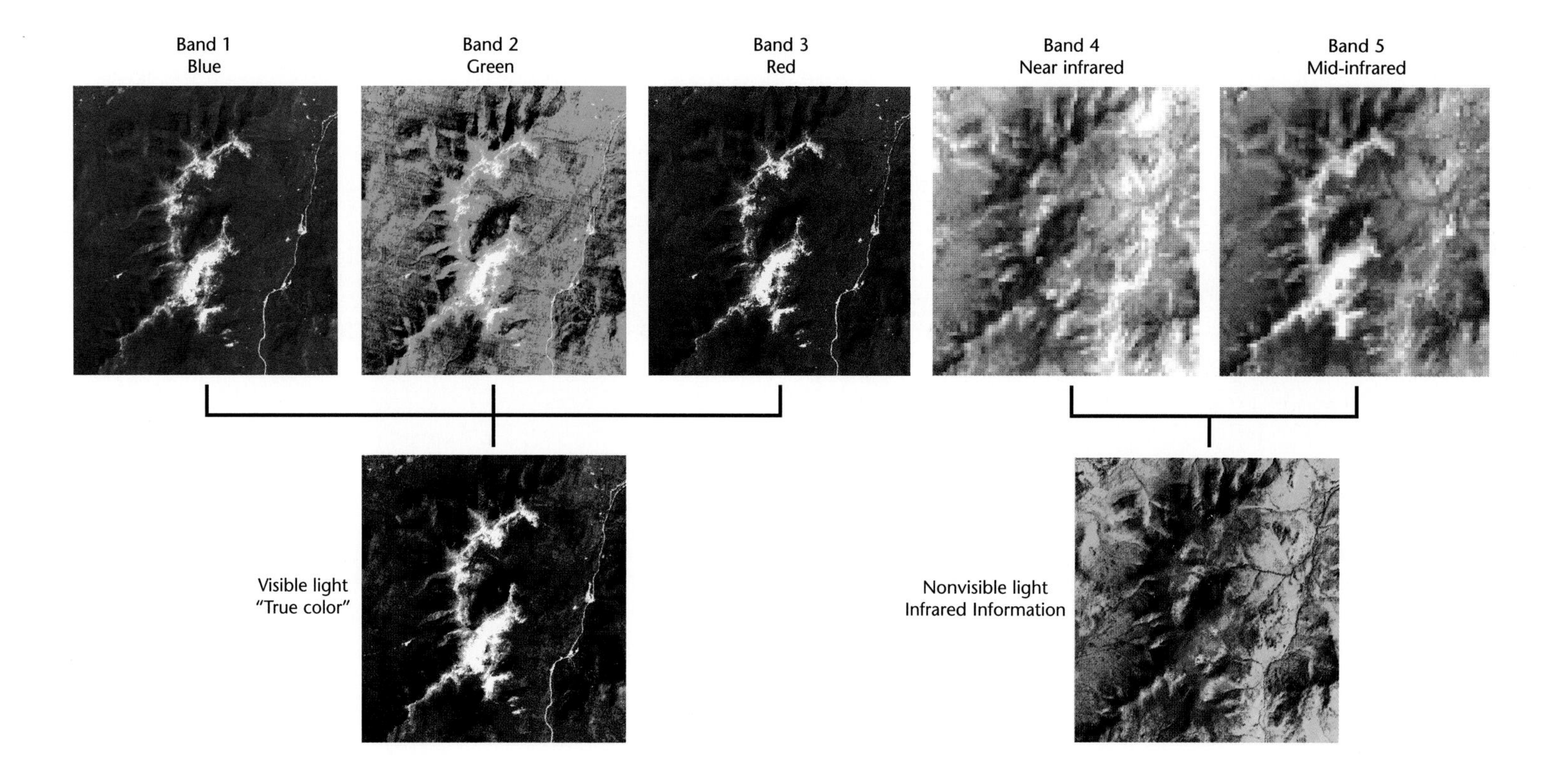

A.5 False-Color Composite

These images illustrate how different wavelengths or band widths are detected by satellite sensors and then blended to create a false-color composite. There are seven bands of data collected by a Landsat satellite that produces Thematic Mapper data. Five of these bands are shown here: blue, green, and red bands from the visible portion of the electromagnetic spectrum, and two from the infrared sector. Different combinations of visible and nonvisible light information can then be blended to create a false-color composite. *Credit: T. Ongaro, Island Institute*

University of Pennsylvania landscape architect Ian McHarg. In his book *Design with Nature* (1969), McHarg described the significance of being able to visualize many aspects of a landscape simultaneously. According to McHarg, this can be accomplished by overlaying clear mylar sheets that display different map data, thereby enabling one to see significant interactions between different characteristics of the landscape. Geographic information systems are designed with this McHargian paradigm in mind, except that overlaying is achieved by coregistering map and image data on a computer.

Implications for the Planet

Although the science of remote sensing has evolved steadily since the first earth-observing satellites launched in 1960, the challenges of interpreting remote sensing data remain large. Satellite images are far more complex than simple photographs. An image based on multispectral data involves measurements of reflected or emitted radiation in several spectral bands. Because human experience is limited to visible sunlight, we have no intuitive knowledge of how features on the earth respond to other forms of electromagnetic radiation. We must rely on experiments, often involving ground-based measurements and airborne instruments, to learn how various features will reflect or emit radiation in different regions of the spectrum.

With satellite images, the possibility of analyzing critical environments anywhere in the world is greatly expanded. Ecologists can study natural and human-induced changes in land use patterns and the global distribution of major biomes. Atmospheric chemists can relate these changes to changes in greenhouse gases, and oceanographers can study physical, chemical, and biological processes at the atmosphere-ocean interface. The ultimate value of remote sensing, we believe, lies in elucidating the linkages between the earth systems—the atmosphere, land, and oceans. Such an ecosystem perspective is substantially enhanced by the availability of space-based measurements and images.

Winston Churchill once said that the farther away from something one gets, the farther into the future one sees. Students in rural Maine schools are certainly looking into the future with their use of satellite images. Whether these students become scientists or not, remote sensing will affect their lives, just as the first photographs of the earth from space changed the lives of generations since 1960. The images sent home by Apollo astronauts had an immediate and lasting impact on how we perceive our watery planet. For the first time we could hold the entire earth in our mind's eye and see it for what it is, an island traveling through space with a finite cargo. By developing an understanding of satellite images, we can all gain an appreciation of the connectedness of ecosystems at all scales.

GLOSSARY

Anadromous: from Greek, literally "up-running"; usually refers to saltwater fish that travel upriver into fresh water to spawn.

Aquifer: a layer of sand, gravel, or *permeable* rock that holds water and allows water to percolate through it.

Benthic: living at or near the bottom of oceans or seas. (See also *Demersal*.)

Biodiversity: the sum of the variability of all living matter, usually referred to in the context of a specific geographic area or ecosystem; e.g., biodiversity of the boreal rocky subtidal zone or of the tropical rain forest. Often measured by the number of species present.

Biome: the sum of all the plants, animals, and other living organisms that form a distinct natural community in a given climatic area.

Boreal: referring to the biogeographical zone south of the Arctic in the cold temperate region in northern latitudes.

Bycatch: non-target species of animals that are often caught up incidentally, often in great volumes, in otter trawls and gill nets; for example, undersized flounders in shrimp nets and harbor porpoises in sink gill nets.

Consumer: an organism that ingests other organisms or food particles; consumers can be parasitic (living off live hosts), herbivorous (living off plants), or carnivorous (living off animals). (See also *Secondary consumer* and *Tertiary consumer.*)

Convective overturn: the vertical replacement of one body of water of a given temperature and density by another; generally cooler water sinks from the surface and is replaced by warmer bottom water, causing an overturn in the water column. The term also applies to a similar exchange of places between air masses in the atmosphere.

Delta: the deposit of earth—mostly sand and gravel—that collects and fans out where a stream or river flows into a larger body of water, such as the sea.

Demersal: living at or near the bottom of the sea; especially said of certain fish such as cod, halibut, and flounder, in contrast to bluefin tuna or mackerel which are *pelagic.*

Detritus: loose material resulting from disintegration; specifically, organic particles or fragments available for reuse within an ecosysytem.

Drumlin: a ridge or oval hill with a smooth summit composed of material deposited by a glacier.

Ecological niche: the set of environmental conditions (temperature, cover, food resources, predators, etc.) in which an organism can be expected to survive.

Ecosystem: a system of ecological relationships in a local environment comprising both organisms and their nonliving environment, intimately linked by a variety of biological, chemical, and physical processes (*Scientific American*).

Ecotone: the boundary or transition zone between two communities or ecosystems.

Estuary: the broad mouth of a river into which the tide surges back and forth, mixing fresh and salt waters.

Eutrophic zone: the area of an aquatic ecosystem that is rich in nutrients such as nitrates and phosphates that promote the growth of algae or flowering plants.

Fault: a zone where rock units have slid by one another, laterally, horizontally, or in any other plane.

Forest association: the community described by a group of dominant plant (tree) species occurring together, such as spruce-fir or northern hardwoods.

Front: the boundary or transition zone between two different masses (of air or water) of differing character, especially temperature.

Groundwater: subsurface water that exists in the pores and crevices of the earth.

Habitat: the place where an organism naturally lives or grows; may be applied to very small areas or to thousands of square miles.

Holdfast: the part of a large marine alga by which it anchors itself to the bottom; analogous to a root in a land plant.

Igneous: one of the three main groups of rocks, formed by volcanic activity or subsurface molten activity. (See also *Metamorphic* and *Sedimentary.*)

Kettle hole: a generally circular hollow or depression in an *outwash plain* or *moraine*, believed to have formed where a large block of subsurface ice has melted.

Metamorphic: referring to rocks that have been altered from their original form as either *igneous* or *sedimentary* rocks.

Moraine: a mass or ridge of earth scraped up by ice and deposited at the edge or end of a glacier. (See also *Recessional moraine* and *Terminal moraine.*)

Non-point-source pollution: the nutrients and contaminants dissolved in water that runs off a general area of land into streams, rivers, and the sea. (See also *Point source.*)

Outwash plain: the plain formed by deposits from a stream or river originating from the melting of glacial ice that are distributed over a considerable area; generally coarser, heavier material is deposited nearer the ice and finer material carried further away.

Pelagic: living in the water column, well above the bottom and some distance from land, as do oceanic fish or birds. (Contrast *Demersal* and *Benthic.*)

Permeable: allowing the passage of liquids or gases.

Phytoplankton: the ensemble of tiny plants that float or drift in marine waters. These tiny plants can produce such dense blooms in the Gulf of Maine that they turn our waters green. Phytoplankton are the base of the food chain on which ultimately most shellfish, fish, birds, and marine mammals depend (the exceptions being those that feed mostly on detritus from benthic plants). (See also *Zooplankton.*)

Pixel: one picture element of a digital image (analogous to one of the dots that comprise a picture in a newspaper or magazine).

Point source: a specific source of pollution at one place; e.g., a pipe that discharges sewage into the water. (See also *Non-point-source pollution.*)

Primary production: the amount of new living matter (biomass) produced by an individual, population, or community; as such, a measure of the productive potential of a community or ecosystem. Phytoplanktonic organisms are primary producers, as are grasses and trees.

Recessional moraine: a moraine formed as a glacier retreated and melted, dropping its load of earth carried underneath.

Rift valley: a valley formed by the sinking of the land between two roughly parallel *faults*, where the spreading of the continents has pulled large blocks of earth apart. Rift valleys marked the first stage in the formation of the Atlantic Ocean.

Secondary consumers: in a food web, those species that depend on chlorophyll-producing plants for food. An example of a secondary consumer in a terrestrial ecosystem is a leaf-eating caterpillar; in a marine ecosystem, zooplanktonic organisms are usually secondary consumers.

Sedimentary: referring to rocks laid down in layers. (See also *Igneous* and *Metamorphic.*)

Spectral band: any defined segment of the electromagnetic spectrum, from very short rays emitted by radioactive substances through the visible colors to long radio waves.

Stochastic: random, unpatterned, variable events.

Terminal moraine: the deposit of sand, gravel, clay, and boulders left behind as a glacier recedes from its leading edge.

Tertiary consumers: organisms that feed on secondary consumers. An example of a tertiary consumer in a terrestrial ecosystem is a bird that feeds on caterpillars; in a marine ecosystem, herring are tertiary consumers. In the Gulf of Maine, the bluefin tuna that feeds on bluefish that feed on mackerel that feed on herring might be considered a sixth-order consumer. Humans who eat bluefin tuna eat at the top of an extremely interdependent, multilayered food web.

Thermocline: a layer within a large body of water that sharply separates parts having different temperatures, so that the change of temperature through that layer is very abrupt.

Tombolo: a double-faced crescent beach that connects an island to the mainland or two islands to each other.

Transpiration: the passage of water and or water vapor through a membrane, such as from a leaf to its surface or surrounding air. Evaporation from the soil and transpiration of vegetation are responsible for the direct return to the atmosphere of more than half the water that falls on the land (Michel Batisse).

Upwelling: a process whereby nutrient-rich waters from the ocean depths rise to the surface; it commonly occurs along continental coastlines.

Watershed: a geologic basin or catchment area that carries precipitation "shed" from the land to the sea.

Zooplankton: the ensemble of small floating animals in an aquatic ecosystem. The zooplankton is composed of a wide range of invertebrates and the larval forms of fish and shellfish. (See also *Phytoplankton.*)

SELECTED READINGS

Apollonio, Spencer. *The Gulf of Maine*. Rockland, Maine: Courier of Maine Books, 1979.

Backus, Richard H. *Georges Bank*. Cambridge, Massachusetts: The MIT Press, 1987.

Bigelow, Henry, and William W. Welsh. "Fishes of the Gulf of Maine." *Bulletin of the United States Bureau of Fisheries*, vol. 40, part 1, 1924. Washington, D.C.: Government Printing Office, 1925.

Conkling, Philip W. *Islands in Time: A Natural and Human History of the Islands of Maine*. Camden, Maine: Downeast Books, 1981.

Edwards, Robert L., Bradford Brown, Marvin Grosslein, and Kenneth Sherman. *Case Concerning Delimitation of the Maritime Boundary in the Gulf of Maine Area*. Volume I, Part A, Annex 1 to the Counter-Memorial: *The Marine Environment of the Gulf of Maine Area.* Submitted by the United States of America to the International Court of Justice, 28 June 1983.

Goode, George Brown. *The Fisheries and Fishery Industries of the United States*. 8 vols. Washington, D.C.: Government Printing Office, 1884–1887.

Harvey, Janice. *Turning the Tide: A Citizens' Action Guide to the Bay of Fundy, Ecology of Coastal Waters*. Fredericton, New Brunswick: Conservation Council of New Brunswick, 1994.

Harvey, Janice. *Voices of the Bay.* Fredericton, New Brunswick: Conservation Council of New Brunswick, 1992.

Irland, Lloyd C. *Wildlands and Woodlots: The Story of New England's Forests*. Hanover, New Hampshire: University Press of New England, 1982.

Kelley, Joseph T., Alice R. Kelley, and Orrin H. Pilkey. *Living with the Coast of Maine*. Durham, North Carolina: Duke University Press, 1989.

Mann, Kenneth H. *Ecology of Coastal Waters: A Systems Approach.* Oxford, England: Blackwell Scientific Publications, 1982.

Mann, Kenneth H., and J. R. N. Lazier. *Dynamics of Marine Ecosystems: Biological-Physical Interactions in the Oceans.* Oxford, England: Blackwell Scientific Publications, 1991.

O'Leary, Wayne M. *The Maine Sea Fisheries, 1830–1890: The Rise and Fall of a Native Industry.* Ann Arbor, Michigan: UMI Dissertation Service, 1994 (degree in 1981).

Thurston, Harry. *Tidal Life: A Natural History of the Bay of Fundy.* Charlotte, Vermont: Camden House Publishing, Inc., 1990.

Van Dusen, Katrina, and Ann Hayden Johnson. *The Gulf of Maine: Sustaining Our Common Heritage.* Augusta: Maine State Planning Office, 1989.

CONTRIBUTORS

John J. H. Albright founded in 1993 The Conservation Group, an ecological consulting company that emphasizes testing and developing the theories of conservation science. He was formerly the director of the Maine Natural Heritage Program from its inception in 1983.

Spencer Apollonio served as the first executive director of the New England Fisheries Management Council and was twice Commissioner of the Maine Department of Marine Resources.

Janet W. Campbell is a research associate professor in the Institute for the Study of Earth, Oceans and Space at the University of New Hampshire. Her research involves the study of ocean biogeochemical processes, in particular the effect of oceanic primary production on global carbon fluxes. She has helped develop remote sensing techniques and applications utilizing airborne and satellite visible and infrared sensors.

Philip W. Conkling is the founder and president of the Island Institute, where he conceived and produced the publication *Island Journal.* He is the author of the book *Islands in Time* (1981), a close study of the natural and human history of the islands along the Maine coast, and of numerous articles on local and global environmental issues. He is the co-designer of GAIA software, the application that displays and analyzes satellite imagery.

Cynthia B. Erickson is a research associate for the Bigelow Laboratory for Ocean Sciences in Boothbay Harbor, Maine. She serves as project coordinator for the Gaia Crossroads Project, a K–12 curriculum development program that explores the value of satellite imagery as an educational resource.

Janice Harvey is former executive director and currently serves as past president of the nonprofit Conservation Council of New Brunswick. A native of Grand Manan Island, she is a founder and current president of the Fundy Community Foundation. She is the author of two books on the Bay of Fundy, *Voices of the Bay* and *Turning the Tide.*

Lloyd C. Irland managed the Spruce Budworm Program at the Maine Department of Conservation from 1976 to 1979 and served as director of the Bureau of Public Lands from 1979 to 1981. He was appointed state economist in 1981, specializing in natural resources, public works, and education policy. In January 1987 he founded The Irland Group, a forestry economics and marketing consulting firm in Augusta, Maine. He is the author of two books and more than 200 articles and bulletins on forestry and natural resources.

Alice R. Kelley has been a consulting geologist on a wide range of topics, an instructor of geology at the University of Maine, and the mother of three children. She was the coauthor with her husband of a book dealing with coastal development problems of the Mississippi River delta. After moving to Maine she again collaborated with her husband on a book concerning unsound development of the Maine coast.

Joseph T. Kelley is the state marine geologist with the Maine Geological Survey. He is based at the University of Maine, where he is an adjunct professor in both the Department of Geological Sciences and the Department of Oceanography.

Kenneth Mann is emeritus scientist at the Bedford Institute of Oceanography, Dartmouth, Nova Scotia, and adjunct professor of biology at Dalhousie University, Halifax. His books include *Ecology of Coastal Waters* (1982) and, with J. R. N. Lazier, *Dynamics of Marine Ecosystems* (1991).

Suzanne Meyer is a landscape designer and graphic artist who served as the principal coordinator for this project.

Annette S. Naegel is a graduate of the Yale School of Forestry and Environmental Studies and is presently the director of science and stewardship at the Island Institute.

David D. Platt is a newspaper editor and reporter, writer, television producer, and freelance journalist. He has worked for the Gulf of Maine Council on the Marine Environment, *Maine Times,* the *Bangor Daily News,* and the Maine Public Broadcasting Network.

Richard Podolsky is the co-designer of GAIA software and has managed a wide variety of geographic information system and remote sensing contracts, including the analysis of Prince William Sound for the Exxon Valdez Restoration Team, the analysis of Joshua Tree National Monument for the National Park Service, and projects for the U.S. Fish and Wildlife Service. He has received research fellowships to study threatened seabirds in Maine, Canada, Hawaii, and the Galapagos Islands.

Harry Thurston is a naturalist, author, and poet who has lived in Nova Scotia all his life. He is the author of *Tidal Life: A Natural History of the Bay of Fundy,* with photos by Stephen Homer.

George M. Woodwell was the founder and remains president and director of the Woods Hole Research Center in Woods Hole, Massachusetts.

Charles S. Yentsch is a senior scientist and founder of the Bigelow Laboratory for Ocean Science in West Boothbay Harbor, Maine. His research interests concern assessments of primary production of the world's oceans. He is a member of NASA's Science Team for the use of ocean color for satellite remote sensing of the oceans.

INDEX

Numbers in italics are figure numbers.